UNDERSTANDING DECISION-MAKING PROCESSES IN AIRLINE OPERATIONS CONTROL

To my dear wife Sue and children Matthew and Jessica

Understanding Decision-making Processes in Airline Operations Control

PETER J. BRUCE
Swinburne University of Technology, Australia

LONDON AND NEW YORK

First published 2011 by Ashgate Publishing

2 Park Square, Milton Park, Abingdon, Oxon OX14 4RN
711 Third Avenue, New York, NY 10017, USA

Routledge is an imprint of the Taylor & Francis Group, an informa business

First issued in paperback 2016

British Library Cataloguing in Publication Data
Bruce, Peter J.
Understanding decision-making processes in airline
operations control.
1. Aeronautics, Commercial--Decision making. 2. Control
rooms. 3. Aeronautics, Commercial--Decision making--
Evaluation. 4. Situational awareness--Evaluation.
I. Title
387.7'4042-dc22

ISBN 978-1-4094-1148-2 (hbk)
ISBN 978-1-138-24792-5 (pbk)

Library of Congress Cataloging-in-Publication Data
Bruce, Peter J.
Understanding decision-making processes in airline operations control / by Peter J. Bruce.
p. cm.
Includes bibliographical references and index.
ISBN 978-1-4094-1148-2 (hardback)
1. Airlines--Management--Decision making. 2. Operations research. I. Title.
HE9780.B78 2011
387.7068'4--dc23

2011016245

Contents

List of Figures

List of Tables

Operational Definitions of Key Terms

The contexts in which the following terms have been applied in this book are described below.

Decision-making style has been defined by Scott and Bruce (1995: 820) as '... the learned, habitual response pattern exhibited by an individual when confronted with a decision situation'.

Decision outcomes are the end result of '... response patterns exhibited by an individual when confronted by a decision situation' (Scott and Bruce 1995: 820).

Disruption '... is an event that prohibits an airline from operating as scheduled' (Rosenberger, Schaefer, Goldsman, Johnson, Kleywegt and Nemhauser 2002).

Domestic OCC refers to an operational control centre which is responsible for airline operations within a country's boundary.

Experience is '... the accumulated memory of past impressions, actions, and achievements' (Khatrih and Ng 2000: 63).

Expertise is the ability of people to use their knowledge to assess a situation, determine if a problem exists, and take appropriate action to resolve the problem (Orasanu and Connolly 1993).

Information Completeness is the process of '... gathering as much information as possible before making a decision' (Endvick 1996: 37).

International OCC refers to an operations control centre which is responsible for airline operations in and beyond a country's boundary.

Situational Awareness is 'the perception of the elements in the environment within a volume of time and space, the comprehension of their meaning, and the projection of their status in the near future' (Endsley 1988: 5).

Time Pressure is the imposition of a deadline that '... may induce a number of different affective states depending upon the extent to which individuals appraise how they can adapt in ways that allow them to maintain their task goals at an acceptable level' (Maule, Hockey and Bdzola 2000: 284).

Tranships are passengers connecting from one flight to another flight.

Preface

The purpose of this book is to provide the reader with some understanding of how and why decisions are made in Airline Operations Control Centres (OCCs) chiefly in response to disruptions. There are currently some excellent books that examine aspects of airline network design and scheduling, and operational disruption management and the reader is referred to these to gain additional information about airline operations centres (e.g., Wu 2010 and Bazagan 2004). Others deal more with economic and financial implications of disruptions as well as the management and systems for information handling (e.g., Rapajic 2009). In some senses, *Understanding Decision-making Processes in Airline Operations Control* may be seen as complementary to these and other texts that deal with airline scheduling and operations in that it attempts to explain ways in which operations controllers go about the decision-making process when confronted with operational disruptions. Where this book can be differentiated from the others is its distinct focus on human aspects that underpin fundamental decision-making processes in OCCs. Hopefully, this will inform the reader of another vital component of disruption management in airlines not previously covered.

The audience for this book is the large cohort of practitioners who are, or have been, directly involved in operational controlling within OCCs, as well as academics and students who are concerned with the teaching or learning of aspects of airline operational control. For the operations practitioner, in particular managers or supervisors in an OCC, as well as those involved with staffing in OCCs, the conclusions and implications for OCC management and airline human resource departments may be useful as a guide for selection and training of new and current operations controllers in airlines. There is no equivalent text that considers the 'human' side of decision making in OCCs, despite anecdotal evidence from airline OCCs suggesting that this is an area that has long been neglected and in need of considerable research. Therefore, this book is likely to be of specific interest to these audiences.

The author has approached the topic area with some degree of credibility as a result of more than 16 years' experience as an operations controller in a major domestic airline. It is this vital experience that has provided sufficient awareness of the problems faced in OCCs on a daily basis in order to undertake the study explained later in the book and then interpret and convey the results in a meaningful way. Indeed, without significant exposure and hands-on experience managing disruptions in an OCC, it would simply not be possible to have the in-depth knowledge required to portray a proficient understanding of the processes involved, nor any idea of the thinking behind solutions to the variety and complexity of problems.

The book is divided into two main parts. The focus of the first part is to provide some underlying background about airline operations and especially the Operations Control Centre and its relevance and importance to airline performance and operational effectiveness. The majority of this part provides theoretical frameworks which will help to form a basis for understanding the study described in the second part.

Chapter One describes the OCC, its functions and areas of responsibility enabling the reader to realise the need for examining aspects of decision-making in OCCs given that the domain has not been overly represented in past work. Chapter Two examines some of the theoretical foundations underpinning decision-making and its importance and relevance in the OCC. Chapter Three introduces two key types of decision-making styles: a rational or classical style and an intuitive style. What becomes interesting in the study is the extent to which controllers in OCCs use one or both of these approaches when managing disruptions. Researchers investigating ways in which people make decisions have also focused on an approach known as naturalistic decision-making (NDM); the study of decision-making in a naturalistic environment and this is also discussed in the context of decision-making in the OCC.

Chapter Four introduces situation awareness in terms of a highly relied upon model, explains the importance of situation awareness in the decision-making process and provides the rationale for linking situation awareness with decision-making. Chapter Five discusses the importance of expertise in aviation and especially in OCCs. Of note is the way in which expertise may be defined and how the performance of novices may be enhanced to produce outcomes expected and normally gained by experts.

Part Two of the book describes a comprehensive study that was conducted to examine decision-making processes of controllers in a number of Operations Control Centres. Chapter Six describes the design of the study and demographic and professional characteristics of the participants who took part. Chapters Seven and Eight present the results of the study separately according to international and domestic simulations. Finally Chapter Nine draws together the main themes and results and gives some direction for further work in this area.

The author offers his sincere thanks to Management in the Operations Control Centres that took part in the study for permitting unfettered access to the personnel, resources, and facilities of their centres. Sincere thanks are also extended to the OCC controllers who willingly participated in the study and probably enjoyed it! Credit must be given to them for coming into the office on their days off, for opening their minds, and for expressing their thoughts freely in the name of research. On a personal level, the author specifically wishes to thank Associate Professor Judy Gray for her expert guidance and to Barrie Colledge, David Johns and John Jones in particular for their assistance in terms of setting up the study and providing valuable feedback. Finally, work like this never achieves its potential without the love and support from family and the author is very appreciative of the continual care and encouragement from Sue, Matthew and Jessica.

Dr Peter J. Bruce
Swinburne Aviation, Melbourne, Australia, September 2011.

PART I
Background and Underlying Theory

Chapter 1
Introduction

The Aviation Industry

Aviation is an exciting, challenging industry that is constantly evolving and responding to innumerable pressures. In this industry, safety is paramount and is controlled rigidly by a comprehensive, regulatory framework of standards, practices and guidelines. Notwithstanding the implications and consequences of a safety mindset, airlines also operate within critical economic margins, necessitating the optimum efficiency of resources and minimisation of costs. A significant characteristic of the aviation industry is the nature of its unpredictability, which results frequently in disruption to schedules, the occurrence of additional costs and unhappy passengers. To respond to these operational disruptions, decisions must be made that help to mitigate the effects of disruptions and decisions yielding optimal results may well be the difference between survival and bankruptcy. Certainly, the consequences of poor operational decisions can be severe.

The Operations Control Centre

No two airlines in the world operate with identical schedules: rather, each airline designs and operates its own network of schedules and it is the management of this network that is central to an airline's operations. At the heart is the Operations Control Centre (OCC) which serves as the airline's nerve centre. The main function of the OCC involves the planning and coordination of the disruption management process to achieve network punctuality and customer service while utilising assets effectively and minimising costs. The OCC domain is a highly complex, dynamic, and fast paced environment in which decisions made by the OCC controllers facilitate disruption recovery. Clearly then, decision-making in this environment is critical, so any improvements to the decision-making process are likely to result in more effective ways in which OCCs can manage disruptions.

The OCC is provided with a planned schedule of flights, reflecting the expectations of booked passengers and setting the parameters within which airline operations are based. The schedule indicates the origin and destination of each flight the airline operates and the days and times at which the flights are planned. The primary scope of responsibility of an OCC lies with the control of flights within a particular period of operation. In international operations, this control may be over several days. However, in domestic operations, control is usually within a calendar day. The chief function of an OCC is to ensure that as far as possible, the operation mirrors the planned schedule. This is performed by monitoring the progress of flights, identifying potential or actual operating problems, and taking corrective actions in response to disruptions.

An Operations Control Centre may vary in terms of location, physical structure, and composition. Depending upon the size of the airline, the centre may be physically located within an airline's head office at a city or airport location. The OCC may consist solely of a group of decision-makers with responsibility for coordinating and controlling aircraft movements. Generally though, the OCC is large enough to include representatives from Pilot and Cabin Attendant Crewing, Engineering, Flight Despatch, airport functions, various commercial and customer service functions, and liaison with Air Traffic Control and Meteorology. A centre incorporating these areas may be known as an Airline Operations Control Centre (AOCC), Integrated Operations Control Centre (IOCC) or a Network Operations Control Centre (NOCC). For the purpose of clarity and consistency throughout this book, the centre will be referred to as an OCC.

Disruption Management in OCCs

There is greater than ever emphasis on maintaining schedule regularity in airline operations, because one of the key measures of an airline's worth and credibility in terms of passenger satisfaction and therefore loyalty is its level of on-time performance. However, airlines constantly face operational disruptions for a variety of reasons such as adverse weather conditions, aircraft maintenance problems, limitations and requirements, air traffic flow and congestion and crewing issues as well as many others. As far as the travelling public is concerned, these disruptions are experienced as flight delays, cancellations or diversions, but part of the disruption recovery may also involve strategies that upgrade or downgrade aircraft sizes, necessitate additional flying to position aircraft for flights or involve moving and rebooking passengers.

Controllers in OCCs need to instigate a series of actions so that the airline recovers from these disruptions promptly. However, the disruptions are often highly complex, in terms of the challenges encountered within individual disruptions, or in terms of the complexities of managing simultaneous disruptions. The necessity to recover from disruptions, often under severe time constraints, calls for rapid and accurate decision-making by controllers in OCCs. Given these challenges, the study described in the second part of this book examines ways in which OCC controllers go about the decision-making process.

Decision-making Process

Figure 1.1 presents the focus of the study described in Part Two in terms of the overall decision-making process. The model indicates that the decision stage is predicated on a number of steps including problem recognition, identification and weighting of decision considerations, and generation of alternatives. The model also indicates that the decision-making process is examined only to the point at which a decision may be made and does not include analyses of actual decisions made. This is because problems in OCCs typically have several solutions, all of which may be feasible and acceptable for resolving a problem. Thus, the focus of the study is predominantly on the individual decision-making processes that enable decisions to be reached, taking into account the extensive range of decision alternatives from which a choice may be made and the limited worth of comparing decision choices.

Previous studies on decision-making have focused on ways in which people make decisions in various situations and the extent to which the effectiveness of decisions can be improved. The need to understand, develop, and enhance decision-making processes has been driven by the growing sophistication and complexity of environments in which decisions often need to be made rapidly with incomplete or inaccurate information. Decision-making in these environments needs to be precise as the consequences of poor decision-making are often very costly.

In the aviation industry the importance of decision-making has been recognised with the focus predominantly on decision-making processes by pilots and air traffic controllers. Not until fairly recently has attention been given to OCCs. Yet, this decision-making environment is extremely intense and the outcomes of decisions critical.

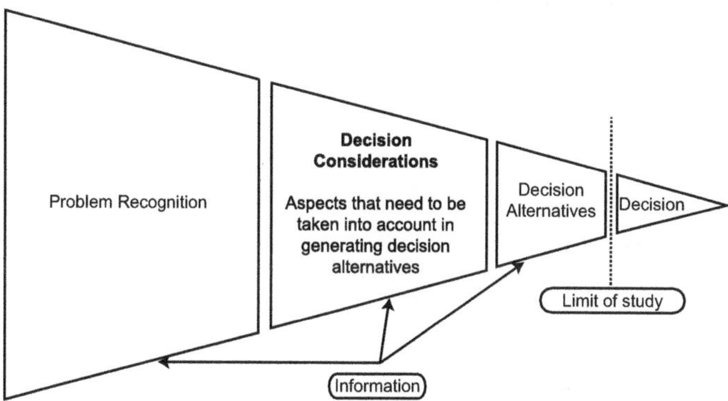

Figure 1.1 Focus of the study in terms of the overall decision-making process

Research that has been undertaken in relation to OCCs has focused on disruption recovery strategies and much of it has involved computer modelling and analysis. Computer-based systems assist controllers by providing information and limited decision-making support. The issue is that the complexity of the problems and need for very rapid decision-making in limited time and with ambiguous or incomplete information limits the usefulness of these systems. In other words, these systems are unable to cope with the intricacies of significant isolated disruptions never-lone the additional complications created by concurrent multiple problems that inevitably occur. Further, the models do not take into account the decision-making processes of individual controllers. However, human intervention in the decision-making process in OCCs is critical as a means of coping with the novelty and complexity of situations. This may help to explain why current practice of OCC controllers is to handle disruption recovery manually rather than rely on computer-based systems. There appear to be very few studies investigating controllers' decision-making processes; yet determining these processes may enable airlines to improve decision-making in OCCs. Part Two of this book considers these processes by examining the relationships among situation awareness, information completeness, experience, expertise, decision considerations, and decision alternatives in OCCs.

Summary

This chapter has provided a brief introduction to the Operations Control Centre of an airline and the key functions performed therein. The main message is the importance placed on efficient disruption management in such a highly complex and constantly changing environment, where human decision-making processes are critical to schedule integrity.

Chapter 2
Decision-making

Introduction

Decision-making underpins many critical activities in the aviation industry. Whereas much research has examined decision-making of pilots and air traffic controllers, very little has considered decision-making processes of controllers in OCCs. Yet, efficient, timely and accurate decision-making in the OCC is vital for its contribution to the operational and commercial success of an airline. This chapter explains decision-making in terms of its importance and relevance to OCCs and investigates variables that may influence the generation of decision alternatives or options, namely situation awareness, information completeness, decision-making styles, experience, expertise, and time.

Definition of Decision-making

A decision has been defined as '... a commitment to a course of action that is intended to produce a satisfying course of action' (Yates, Veinott and Patalano 2003: 15) and decision-making as '... intentional and reflective choice in response to perceived needs' (Kleindorfer, Kunreuther and Schoemaker 1993: 3). In an airline OCC, decision-making involves a process of monitoring the progress of flights and reacting to operational disruptions in order to improve on-time performance, reduce operational costs, and increase customer satisfaction. Clearly, decision-making in an OCC environment is critical to all aspects of airline operations.

Importance of Decision-making

The decision-making process has long been regarded as one of the most important and pervasive of activities as it affects many day to day endeavours. As a result, extensive studies about decision-making have been conducted stemming from a desire to understand how people make decisions in various situations and how the effectiveness of decision-making may be improved. The growing sophistication and complexity of decision environments has created a greater need for understanding, developing, and supporting effective decision-making. Of course, part of this push for greater understanding is the need to gain a greater awareness of the consequences of *poor* decisions.

In response to these needs, considerable research has examined decision-making and its relationships with uncertainty, ambiguity and risk, choice, judgment, experience, and expertise. A common goal evident in much of this work has been a focus on improving decision-making processes and identifying factors and dimensions that are related to those processes in order to improve decision outcomes. The research has also led to a greater understanding of decision processes employed by experts. In many disciplines or domains, decision-making is recognised as being highly complex and sensitive to task and domain factors such as expertise, time, and information. These aspects also characterise decision-making in aviation.

Decision-making in Aviation

In aviation, considerable research has examined decision-making processes with regard to aspects such as tourism and aviation policy, airline profitability, airline service quality performance, revenue management, and aviation business services marketing. However, studies that have examined *decision-making* in aviation have focused predominantly on decision-making by commercial and military pilots and air traffic controllers and much of it has emphasised decision-making in relation to safety aspects. This is because pilots and air traffic controllers need to make precise, informed decisions often with ambiguous information and often with limited time in an environment in which the consequences of poor decision-making may be catastrophic. Despite the amount of research in aviation decision-making, limited studies appear to have been conducted in airline OCCs and the research that has been conducted has yet to consider decision-making processes of controllers in these OCCs.

Decision-making for Disruption Management in OCCs

Decision-making in OCCs is made difficult by the numerous operational disruptions faced by airlines. These disruptions result from factors such as inclement weather, maintenance breakdowns, pilot and cabin crew limitations, passenger handling problems, air traffic control clearances, and airport congestion and are most visibly evident in delays, cancellations, and diversions. Although poor decisions of controllers in OCCs may not culminate in safety incidents, sound decision-making is critical to achieve desired operational outcomes such as on-time performance, customer satisfaction, economic operations, and rapid recovery from operational disruptions.

Past research in airline operations has investigated planning activities such as schedule optimisation and fleet assignment (e.g., Ball 2003). Some of this research has resulted in the development of models that assist airlines to plan schedules for profit maximisation and optimum carriage of passengers (e.g., Soumis, Ferland and Rousseau 1980). Other models have augmented these by incorporating fleet assignment (e.g., Lohatepanont and Barnhart 2004), maintenance, and/or crew assignment (e.g., Lettovsky, Johnson and Nemhauser 2000).

Managing the operations on a current day basis though is quite different from designing optimal schedules. The controllers in an airline's OCC are charged with this task and when operational problems occur, they need to make decisions quickly and effectively in order to resolve the disruption and minimise passenger inconvenience and cost to the airline. So, a growing body of research is beginning to focus more extensively on decision-making in relation to disruption management, often employing mathematical modelling techniques. For example, researchers such as Clarke (1997) used linear programming and network flow theory to assist airline schedule recovery following disruption and Cao and Kanafani (2000) developed an algorithmic model to analyse the value to the airline of runway availability and the effect on the airline's schedules. The problem is that to date, none of the models has been able to cope with the complexities of multiple, simultaneous disruptions commonly experienced by airlines.

Literature examining disruption recovery strategies is quite extensive. Much of the disruption research has focused on aircraft routing, aircraft maintenance, and crew scheduling (e.g., Cohn and Barnhart 2003, Lederer and Nambimadom 1998). Some of the disruption recovery strategies have included optimisation models (e.g., Abdelghany, Shah, Raina and Abdelghany 2004, Rosenberger, Johnson and Nemhauser 2003), or have examined the economic effects of disruptions (e.g., Janic 2005). In order to reduce the complexities of these models and enable them to work, assumptions have often been made in order to limit the parameters included in the models. For example, models developed by Talluri (1996) and Rosenberger, Johnson and Nemhauser (2003) did not take into account and therefore were greatly limited by the inability to substitute different *aircraft types* to solve disruptions, despite typical airline fleets consisting of a variety of aircraft types. The work of Yan and Yang (1996, 2002) and Yan and Tu (1997) resulted in the development of several models but again none of these took into account the ability to change aircraft types to help solve problems; nor did these models consider crew and maintenance parameters. Sriram and Haghani (2003) restricted their study to domestic airline operations and considered only limited maintenance parameters and Rosenberger et. al. (2002) worked with single disruptions.

Despite this extensive research in disruption management, no single optimisation model has yet been developed to address the task required to solve complex operational problems. According to Barnhart, Belobaba and Odoni (2003: 369), '... the problem's unmanageable size and complexity has resulted in the decomposition of the overall problem into a set of sub-problems ...' These sub-problems typically include network and schedule design, fleet resource assignment, maintenance requirements, and crew scheduling. Further, it is apparent that computer-based tools cannot cope with the limited decision-making times required in disruption management.

These are key points and suggest that regardless of computer-based intervention, the decision-making processes of controllers in OCCs appear to rely largely on the human's ability to identify, assess and carry out actions to solve disruptions, most likely using a combination of intuition and experience. Indeed, one researcher (Lettovsky 1997) asserts that airline OCCs mostly recover from disruptions manually.

The trouble is that identifying how people use intuition makes airline disruption recovery difficult to comprehend, which may help to explain the deficiencies in the models developed so far. In other words, the research based on developing models to solve operational problems in the industry has had limited success because it has not taken into account the underlying human decision-making processes required to solve problems in such a complex environment.

The complexity is emphasised by the endless variety, uniqueness and combinations of problems. By limiting criteria to single disruptions or simple aircraft pattern changes, or by including little or no crewing or maintenance parameters, the models will never replicate true disruption management. Yet, incorporating all of these parameters, if it could be done, would require so much computer power and extensive calculation time that the optimum decision point would quickly pass. And this is if circumstances remained steady during this whole process: in domestic operations especially, not a reality! It is the work of these and other researchers that has led, in part, to this book as very little research appears to have examined individual decision-making processes leaving a gap between what Clausen, Larsen and Larsen (2005: 1) describe as '... the reality faced in operations control and the decision support offered by the commercial information technology systems targeting the recovery process'.

Decision-making and Problem Solving

According to Tversky and Kahneman (1985: 25), a decision problem is defined by '... the acts or options among which one must choose, the possible outcomes or consequences of those acts, and the contingencies of conditional probabilities that relate outcomes to acts'. Problem solving on the other hand is concerned mainly with the search for possible options to achieve the desired goal (Baron 1994). Bainbridge (1999) suggests that problem solving is a term used to describe the mental processes used by someone in an unfamiliar situation about which the person has insufficient knowledge, or for which the method of dealing with the problem is inadequate. A problem comes about when there is a gap between an initial situation and a desired situation, with no obvious way to accomplish change to reach the desired situation (Klahr and Simon 2001). The gap is caused either by dissatisfaction with a current situation, or by attraction to alternative possibilities and opportunities.

The terms *decision-making* and *problem solving* have been used variously in the literature and at times, interchangeably. For example, decision-making has been seen both as contributory to problem solving and as a response to problems. The terms have also been clearly distinguished. For example, decisions can be made in the absence of a problem and a problem can be solved without making a decision. Decision-making may include the steps of finding and defining a problem and generating and evaluating solutions or may be a problem solving process with the decision as the solution.

Despite the debate that surrounds the use of the terms decision-making and problem solving, decision-making in this book is considered to be a process involving the generation and evaluation of possible solutions to identified problems, with a view to implementing the optimum solution. In this context, decision-making is regarded as a component of problem solving. In airline OCCs, operational problems disrupt planned schedules and require decision-making by controllers to minimise impact of the disruptions. Hence, decision-making is conceptualised as a means for solving operational problems.

Decision-making, Judgment and Reasoning

Substantial research has investigated the relationships among decision-making, judgment and reasoning. The body of literature is immense and the interested reader is referred to other sources as a comprehensive discussion is beyond the purpose of this book. In Part Two of this book, the role of judgment and reasoning in decision-making is alluded to in terms of the way in which individuals tacitly evaluate and analyse situations, then generate alternative actions from which a preferred option is selected for implementation.

Decision-making Under Time Constraints

Much interest has been shown in relation to time available for decision-making and has distinguished between time constraints and time pressure. This interest has come about due largely to the impact of limited time for the decision-making process which may result in sub-optimal decision processes (Ariely and Zakay 2001). Time constraints have been described as specific allotments of time for decision-making and occur if a deadline exists (Ordonez and Benson 1997). In contrast, time pressure arises as a result of perceiving time allotments as insufficient for decision-making (Rastegary and Landy 1993). To minimise confusion and provide consistency, the term used in this book will be time constraints to denote situations where time for decision-making by OCC controllers is limited.

Where time for decision-making is limited, decision-makers often cannot find a solution to a problem, or cannot find the most effective result. As a result, they may *buy* time in order to assess the situation comprehensively and defer any decision, and may only consider a limited number of alternatives in decision-making. Thus, examining the influence of time constraints is important for decision-making because having limited time appears to change the way people make decisions.

Research has examined the influence of time in terms of situation awareness (e.g., Endsley, Bolté and Jones 2003), information completeness (e.g., Ahituv, Igbaria and Sella 1998), and decision-making styles (e.g., Johnson, Payne and Bettman 1993). Attention has been drawn to ascertaining appropriate information when decision-makers are subject to time constraints and in these situations, the amount of time available for decision-making limits the amount and quality of information necessary for making effective decisions. The response of decision-makers in these circumstances may be to process information faster, filter specific information or moderate their decision processes. The research on time constraints has focused predominantly on domains such as commercial and military pilots and air traffic control where time for decision-making is very limited and the decision environment is highly complex and dynamic. However, very little research appears to have examined time constraints in relation to decision-making processes of controllers in OCCs; yet it is likely that time constraints may influence decision-making processes of controllers given the complexity of the OCC environment. In Part Two, decision-making under different time constraints is examined by considering disruption management in both international and domestic environments. As experienced OCC controllers would recognise, disruption management in each of these environments requires quite unique techniques.

Generation and Evaluation of Decision Alternatives

A fundamental step in the decision-making process is the generation and evaluation of alternatives, from which a choice may be made to select some course of action. Although the terms *decision options* and *decision alternatives* are often used interchangeably to describe a range of possible choices, decision 'alternatives' will be used for the sake of clarity and consistency. According to Janney and Dess (2004), the evaluation of alternatives implies a cost-benefit process where possible solutions are weighed against each other to determine which one is most beneficial to the organisation. A value is assigned to each aspect of each criterion and the alternative with the highest aggregate value is chosen. However, this process takes time, incurs costs, and requires 'effort' and is not an appropriate method for decision-making in complex situations under time constraints.

A major determination in evaluating alternatives is ascertaining whether the decision-maker seeks the best possible solution or is satisfied with some lesser solution. Simon (1976) calls this 'satisficing'; that is, the decision-maker looks for a course of action that is satisfactory or 'good enough'. Thus, the first alternative that meets the criteria is likely to be selected. This limited result is due to the decision-maker's recognition that reaching the optimum decision is too time consuming, information is unknown, or the cost of delaying the decision or finding a better alternative is too great. It also assumes that the decision-maker is aware of other alternatives. In airline OCCs, decision-making is highly complex as a result of the number of stakeholders with vested interests in decision outcomes.

Decision-makers need to devise several alternative strategies for handling operational disruptions as well as develop a number of contingency plans should alternatives be rejected (e.g, inability to provide crews) or circumstances change significantly (e.g, further maintenance work for rectification). The extent to which decision-makers in OCCs are able to generate and evaluate alternative courses of action in relation to disruptions in explored in Part Two.

Summary

The purpose of this chapter was to define and explain the importance of decision-making and in particular the relevance of decision-making for disruption management in OCCs. A focus of the chapter was the consideration of having to make decisions in short time frames and the need often to find solutions to problems without necessarily pursuing or even identifying all options.

Chapter 3
Decision-making Styles

Introduction

Decision-making style has been defined as '... the learned, habitual response pattern exhibited by an individual when confronted with a decision situation' (Scott and Bruce 1995: 820). It has also been suggested that decision-making styles lie along a cognitive continuum from analytical to intuitive extremes (Hammond 1993); an interpretation which suggests that the use of one approach precludes the other. In contrast, other arguments contend that intuition is combined with rational thought to expedite decision-making (Sadler-Smith and Shefy 2004b), and that managers in the field use both types of decision-making (Simon 1987). It seems that there is some dilemma as to the most appropriate decision-making style to use in situations. Stewart (2006) suggests that the right approach is to use the insights of both styles. Studies of decision-making behaviour have shown that selection of a decision strategy depends on factors such as complexity and time pressure (e.g., Janis and Mann 1977). Anecdotal evidence from OCC managers and controllers suggests that controllers in an OCC environment may rely on various combinations of rational and intuitive decision-making approaches at times. Therefore, examining the decision-making styles of controllers in OCCs should lead to a greater understanding of the extent to which controllers use both rational and intuitive decision-making styles.

Rational Decision-making Style

The expressions classical, analytical, scientific and rational have all been used at different times and in various circumstances to refer to decision-making. Most commonly though, the term *rational decision-making* has been applied, describing decision-making as a logical, systematic process of analysis that occurs in a series of steps (e.g., Beach and Lipshitz 1993, Dastani, Hulstijn and van de Torre 2005). Despite its application, Janis (1989) suggests that the term *rational* decision-making may be ambiguous and should not be used. Part of this ambiguity arises from perceiving decision-making as 'not rational' which, in Janis' view, connotes 'irrational' or 'unrealistic' behaviour. However, the term's overwhelming use in many studies suggests that the word *rational* is quite appropriate and acceptable. To minimise confusion between this and other expressions for the purpose of this book, the term *rational decision-making* is used consistently to refer to a logical, step-by-step, systematic approach to decision-making.

Several advantages of using a rational decision-making approach have been identified. For example, Sadler-Smith and Shefy (2007) attest to the superiority of the rational style for decision-making and problem solving as it is stable, predictable, and able to be generalised. Other arguments contend that decision-makers using this approach optimise the use of information in a deliberate, thoughtful manner and increase the generation and comparison of many alternatives (e.g., Corner, Buchanan and Henig 2001). According to Bonabeau (2003), the use of a rational approach seems appropriate for weighing information and evaluating the alternatives. However, this approach assumes that suitable alternatives can be identified and there is capacity to '... combine the information in an objectively optimal manner' (Lord and Maher 1993: 21).

A major strength of the rational approach is the ability of the decision-maker to use a process that follows a number of steps without having prior knowledge or experience upon which to draw. This seems appropriate for circumstances in which there is abundant time for decision-making. However, other arguments suggest that the rational approach may not work in situations that are very complex (Vaughan 1979) or have time-constraints (Wong 2000).

Rational Decision-making Models

A number of simple models have been developed to support a rational approach to decision-making (see Figure 3.1), enabling complex processes to be simplified conceptually. Despite differences in the number of stages and variations in the emphasis given to different stages of decision-making, it is evident that a large degree of commonality exists among these models. They refer variously to a problem recognition stage, a generation and analysis of alternatives stage which calls for the provision of information, and a choice stage.

In aviation, a number of prescriptive models have been developed to aid decision-making. Many of these models apply specifically to pilot training and typically use acronyms to prompt sequential activities in the decision-making process. For example, the 'DECIDE' model specifies several steps: *D*etect problem, *E*stimate consequences, *C*hoose, *I*dentify, *D*o, and *E*valuate (Benner 1975). Mather's (1989) 'PASS' model includes the steps: *P*roblem identification, *A*cquisition of information, *S*trategy survey, and *S*election of strategy. Hörmann's (1995) 'FORDEC' model is based on six stages: *F*acts, *O*pinions, *R*isks, *D*ecision, *E*xecution, and *C*heck. Although these models and the models summarised in Figure 3.1 emphasise several stages that occur in a rational decision-making approach, none of the models prescribes explicitly the considerations of factors that decision-makers need to take into account in evaluating alternative courses of action. However, it would seem important to identify these considerations clearly in the decision-making process in order to optimise solutions. This appears to highlight a weakness common to rational decision-making models. While further elaboration of the models may lead to a better understanding of the rational decision-making approach, other limitations of the rational decision-making approach are also evident.

Beach and Mitchell (1978)	Hogarth (1987)	Janis and Mann (1977)	Mullin and Roth (1991)	Nutt (1999)	Simon (1960)
Recognising problem	Structuring problem	Appraising	Recognising problem	Diagnosis – Signals, Information gathering, Determining need for action	Intelligence – identifying problem and collecting information
Evaluating task	Assessing consequences	Surveying alternatives	Generating alternative choices		
Selecting strategy	Assessing uncertainties				Design – planning for alternatives
Processing information	Evaluating alternatives	Weighing alternatives	Evaluating choices	Action – Establishing direction, Identifying options, Developing plan, Evaluating, Implementing, Review	
	Conducting sensitivity analysis				
Implementing strategy	Gathering information	Deliberating decision	Assessing and modifying decision		Choice – selecting and monitoring solution
Choosing	Choosing an alternative	Adhering to decision			

Figure 3.1 Rational decision-making models

Limitations of Rational Decision-making

Several researchers have criticised the rational decision-making approach due to its limited success in real-life situations (e.g., Brehmer 1999, Klein 1993, Shafir, Simonson and Tversky 1997). According to Nutt (1989, 1999), the rational method provides a logical foundation for decision-making, but fails to ensure good outcomes. This may be because with the step-by-step approach evident in rational decision-making, decision-makers focus extensively on the *sequence of steps* rather than the tasks needing to be performed. However, in many domains such as fire-fighting, medicine, offshore gas installations, and emergency services, decision-making cannot be explained in terms of a step-by-step approach, as the situations are complex and circumstances change dramatically. As a result, decision-makers do not have time to compare a number of alternatives before making a decision. Klein (2001) suggests that in addition, they may have insufficient information on which to base decisions, or have too much information to analyse. In these situations, rather than collecting and evaluating information prior to decision-making, people often make decisions quickly and support the decisions afterwards. Accordingly, there may be little similarity between what rational theory prescribes and how decisions actually are made. Watson (1992) suggests that a rational approach is only used in ten per cent of decision events; in the other 90 per cent, decision-makers use intuitive decision-making.

Rapidly changing circumstances are characteristic of the complex, dynamic environment of an airline's OCC. Thus, decision-making in this domain may not follow a rational approach due largely to the time constraints under which decisions must be made. This has also been evident in plenty of studies that have examined decision-making in emergency or time-constrained situations (e.g., the work of Flin, Stewart and Slaven (1996) and Flin, Salas, Strub and Martin (1997)). Therefore, it is likely that a rational decision-making approach in this environment may only partly explain how controllers in OCCs make decisions.

Part of the criticism of a rational decision-making approach can be attributed to problem structure. In most cases, rational decision theory has been concerned with well defined problems. However, decision-makers usually do not face well-defined problems in turbulent environments in which decision-making is complex and difficult to predict and in which there are a large number of conflicting goals. Thus, in these circumstances, rational decision-making may not be adequate for solving problems, and other methods of decision-making may be more successful. In OCCs, it appears that alternative decision-making approaches such as the use of intuition may be more appropriate at times than rational decision-making for dealing with complex, ill-structured problems.

Intuitive Decision-making Style

Intuition has been defined as '... a capacity for attaining direct knowledge or understanding without the apparent intrusion of rational thought or logical interference' (Sadler-Smith and Shefy 2004a: 77). A number of studies have considered intuitive decision-making to explain decision-making processes when a rational approach is inappropriate (e.g., Buchanan and O'Connell 2006, Leybourne and Sadler-Smith 2006), but the problem with studying intuition is that it is difficult to examine ways in which decision-makers arrive at solutions intuitively without necessarily being able to report how they attained the result.

Intuitive decision-making has been described in many ways. For example, Agor (1984: xii) suggests that intuitive decision-making is '... an uncanny ability to sense intuitively how to make the right decision at the right time – even when information on which to base important decisions was incomplete or totally inadequate'. Intuition may also refer to having a 'hunch' or 'gut-feeling' and especially having a feeling as to what might work or might not work, or sensing what is right and what is wrong (Hedlund and Sternberg 2000, McLucas 2003). This intuitive approach stems from tacit, unspoken knowledge which is gained from experience, enabling decision-makers to take action without first having to think through the action.

Decision-makers who decide intuitively often cannot explain how their decisions are reached. They do not follow the deliberate, sequential steps involved in rational decision-making. Instead, they report having a feeling for what is required, consider many alternatives concurrently, and rely on their intuition to make sense of a chaotic situation; an approach that offers a fast and accurate means of decision-making (Dane and Pratt 2007). In this regard, intuitive decision-making is quite distinguishable from rational decision-making. Importantly, intuition is not regarded as 'wild guessing'. According to Dreyfus and Dreyfus (1986: 29) intuition is '... the sort of ability we all use all the time as we go about our everyday tasks'. Further, a judgment is called intuitive '... if it is reached by an informal and unstructured mode of reasoning, without the use of analytical methods or deliberate calculation' (Kahneman and Tversky 1982: 124). In light of these viewpoints, the use of intuition may be beneficial in situations in which circumstances constantly change and where limited time prevents a more analytical approach to problem solving. These situations typify the environment encountered in airline OCCs and consequently the role of intuition is examined in the study forming Part Two of this book.

Extensive research has considered the relationship between intuition and experience, arguing that experience forms the basis of intuitive decision-making (e.g., Bonabeau 2003, Cesna and Mosier 2005, Klein and Weick 2000). In Simon's (1986) view, acquiring a sufficient level of experience takes about ten years. As OCCs are domains characterised by these aspects, it is reasonable to expect that controllers may use intuitive approaches in decision-making and it is also likely that more experienced controllers may use an intuitive approach to a greater extent than less experienced controllers. Thus, the study in Part Two examines the use of intuition in OCCs and in particular the relationship between experience and intuition.

Intuition and Heuristics

As part of an intuitive approach to decision-making, individuals may employ heuristics to reduce processing demands '... by enabling simple rules to replace detailed representations and lengthy search procedures ...' (Roberts 2004: 265). These shortcuts or 'rules of thumb' enable decision-makers to simplify situations by selecting some aspects of a situation while ignoring others. Several researchers allude to the importance of heuristics which enable individuals to simplify or disregard information due to their limited cognitive abilities and their inability to interpret vast amounts of data (e.g., Marsh, Todd and Gigerenzer 2004, Payne, Bettman and Johnson 1990). Other work has emphasised the usefulness of heuristics when a routine and exhaustive check cannot be made due to time or knowledge constraints (e.g., Johnson, Payne and Bettman 1993, Simon 1982), or when there is a need to make speedy and accurate decisions without the need for lengthy, analytical processes (Roberts 2004, Todd and Gigerenzer 2000). Tversky and Kahneman (1982) suggest that decision-makers use a limited number of heuristic principles to aid in judgment. However, to prevent distortion of a decision outcome, heuristics must rely on correct and relevant information. Heuristics may be valuable by saving time and effort, but they may lead to biases and may not necessarily provide the optimum solution with the information available at the time of the decision.

In OCCs, heuristics have been used in mathematical simulations to develop disruption recovery models (e.g., Lettovsky 1997, Rosenberger, et. al. 2002). However, these models have shortcomings as the heuristics used simplify the mathematical processes employed by limiting the conditions that need to be satisfied. For example, restrictions are placed on the numbers and types of aircraft changes, and patterns of flight crew rosters. Consequently, models based on heuristics have so far failed to provide a reliable means for solving complex problems in OCCs. While the study recognises that controllers may use particular rules of thumb in the decision-making process, there was no specific examination of heuristics in the study described in Part Two. However, this is an area for future research.

Naturalistic Decision-making

Past research has criticised the inadequacies of a rational decision-making approach for explaining the way in which decisions are made in natural contexts (Klein 1997) and in particular with regard to its failure to address the presence of uncertainty common in natural surroundings (Sonenshein 2007). Sonenshein (2007: 1024) defines uncertainty as '... a lack of information that makes constructing a plausible interpretation about a situation difficult'. According to Cannon-Bowers, Salas and Pruitt (1996: 195), '... classical decision-making research was seen as focusing on sterile, contrived decision-making situations, with results that were of little consequence to real-world decision-makers'. This inadequacy to account for real-world decision-making has led to the development of naturalistic decision-making. Gary Klein is considered a leading researcher in naturalistic decision-making (NDM), describing it as describing the way people use their experience to make decisions in natural settings (Klein 1997). The aim of NDM is to obtain a higher quality of decision by means of comprehending decision-making processes that occur in natural contexts rather than contrived circumstances.

The rational decision-making process is based typically on the assumption that alternatives, preferences, and outcomes are known and evaluated in advance. However, this view is flawed in circumstances in which the range of alternative solutions that may be generated changes. Note that in the study discussed in Part Two, no assumptions are made with regard to the possible alternatives. Other studies (e.g., Beach and Lipshitz 1993) argue that in naturalistic settings, traditional models of decision-making that focus on evaluating and choosing courses of action do not apply and in these circumstances, decision-makers rarely use analytical methods. Further, rational decision-making is considered inappropriate to account for the experience of decision-makers, the complexity of tasks, and the demands of the naturalistic environment such as time pressure and ambiguous information (Crichton and Flin 2002). These criticisms of the rational approach underpin the need to study decision-making in a more natural state.

Characteristics of NDM

Orasanu and Connolly (1993) outline the key characteristics of naturalistic decision settings as being: ill structured problems, uncertain, dynamic environments, shifting, ill-defined, or competing goals, action/feedback loops, time stress, high stakes, multiple players, and organisational goals and norms. In these settings, there are significant consequences for incorrect decisions and actions (Klein and Klinger 1991). Further, decision-makers have little control over the occurrence of events and cannot necessarily control the rate of decision-making or the timing of decisions. NDM has been examined in many dynamic environments such as fire fighting, the military, nuclear power plants, nursing, emergency management, transport and sport. The NDM approach used in various studies has enabled researchers to collect data as a result of verbal exchanges and by observation of actions as they take place in the field. Accordingly, the NDM approach is seen as a valuable method for examining decision-making processes as it '... incorporates the interplay between task, person, and environmental factors' (Currey and Botti 2003: 207).

Methods of examining decision-making processes in naturalistic settings often have required a qualitative approach that enables the researcher to gather data from individuals by observation and interview. As a means of studying real-time contexts, the NDM perspective has been criticised for using small sample sizes and for relying largely on participants' verbal protocols. The criticisms have also been largely refuted with counter reasoning arguing that the sample size is likely to be limited due to the extremely time-consuming research process involved with a larger sample, and that the verbal protocols used do indeed provide a rich source of comprehensive, qualitative data. Therefore, it appears that studying decision-making from an NDM perspective is likely to yield more meaningful outcomes than other methods. Airline OCCs are complex, dynamic environments, with additional levels of complexity brought about by the frequent occurrence of multiple, simultaneous disruptions. Thus, examining decision-making within an NDM framework may assist in developing an understanding of controllers' decision-making processes in airline OCCs.

NDM Models

Several models of naturalistic decision-making have been described in previous research. For example, Lipshitz (1993) reviewed nine NDM models that focus on real-world decision-making. The main emphases of these models is their reliance on recognising and assessing situations, forming mental pictures, and utilising intuitive rather than rational approaches in decision-making. In a review of four further models of NDM, a conclusion drawn by Wong (2000) was that decision-makers place great importance in regard to reliance on *pattern-recognition* as a means of evaluating courses of action. According to Klein (2003), a pattern is a set of cues that usually chunk together such that if a few of the cues are identified, other cues are also likely to be found. The recognition of patterns among the flight displays used by OCC controllers may be important in terms of the study as it may provide insight into their decision-making processes. However, the patterns are complex and diverse and as such may be difficult to examine given the individual nature of each disruption.

Klein is recognised as a key researcher in naturalistic decision-making and his recognition-primed decision-making (RPD) model is acknowledged as the dominant naturalistic decision-making model. Klein's model describes how expert decision-makers assess a situation using their experience, from which they are able to generate and evaluate a satisfactory alternative, without the need for comparative evaluations. According to Hedlund and Sternberg (2000), this approach is more representative of the way in which decisions are made in the real world than analytical models. The levels of experience enable decision-makers to recognise patterns from which they can identify achievable goals, relevant cues, and the actions most likely to succeed (Kaempf, Klein, Thorsden and Wolf 1996). In terms of the study, the RPD model may provide insight into ways that expert controllers assess disruption situations and the extent to which they generate and evaluate suitable alternatives. Klein suggests that the model does not account for decision-making by novices who need to be trained to recognise situations, but then need to acquire experience to enable them to conduct mental simulations. Figure 3.2 presents a grossly simplified adaptation of Klein's Recognition Primed Decision (RPD) model of decision-making. The source of the complete model is Klein (1993) in the reference list at the end of this book.

The evaluative stage in decision-making presented in the RPD model is important in terms of the study. The evaluation is conducted by simulating mentally the course of action to see if it will work, and then looking for any unintended consequences that might be unacceptable. However, the mental simulation process does not indicate explicitly the considerations of factors that decision-makers take into account during this process. The exclusion of decision considerations has been questioned earlier in this chapter with respect to the rational decision-making approach. Notably, decision considerations should not be confused with *cues* which are referred to variously as sources of information. Klein refers to cues as an aspect of the recognition stage in the RPD model. In terms of the study described in this book, cues and decision considerations are clearly differentiated. For example, a cue in the OCC may be an advice from say, Maintenance, informing a controller of a technical problem with an aircraft and an anticipated delayed flight. In contrast, a decision consideration that the controller may need to take into account may be to do with recognising the likelihood of influencing a maintenance service *as a result* of the delayed flight. In OCCs, controllers involved in disruption problem solving allow for many considerations such as crew availability, maintenance limitations, airport restrictions, curfews and many others, any of which may influence their choice of decisions. Therefore, it would seem important that considerations form an explicit component of the decision-making process. Accordingly, the study examines the extent to which controllers take considerations into account during this process.

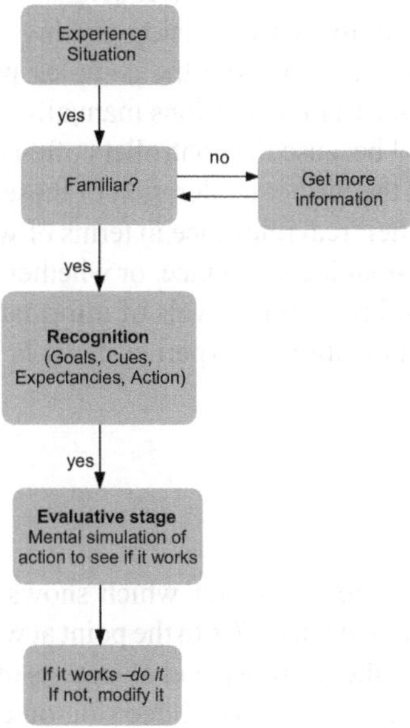

Figure 3.2 Simplified adaptation of Klein's Recognition Primed Decision (RPD) model of decision-making

Expert Systems

Other studies have investigated the importance of human-systems interaction (e.g., expert systems) as a means of helping decision-makers reach outcomes (e.g., Barthelemy, Bisdorff and Coppin 2002, Chu and Spires 2001). Expert systems have been defined as '... knowledge-based software tools or decision aids, intended to assist experts to reach outcomes (Hoffman, Shadbolt, Burton and Klein 1995: 130). The purpose of expert systems may not be to solve a problem, but to formulate the problem in an efficient way or provide storage of expert knowledge (Balch, Schrader and Ruan 2007) such that specific recommendations can be provided. They should assist decision-making by asking relevant questions and explaining the reasons for adopting certain actions. Thus, these systems should be regarded as decision *aids* rather than a means of replacing the decision-maker and as such are likely to lead to a better decision than the use of either method in isolation. Other criticisms of expert systems suggest that the use of them leads to an underestimation of human expertise for problem solving (e.g., Ahituv and Neumann 1987) and that further improvements to decision-making require a greater level of understanding as to how humans process information to aid the decision-making process (Barthelemy et. al. 2002).

In aviation, expert systems are certainly employed as real-time decision support tools to help the management of airlines (Abdelghany, Abdelghany and Ekollu 2008), but despite attempts to assist decision-making in OCCs, it appears that none of these systems can yet cope with the complex problem solving required. Thus, OCCs largely recover from disruptions manually, making airline disruption recovery difficult to model because the controllers often act on their intuition. The use of expert systems and the further exploration of these in OCCs requires a great deal of work to establish their real relevance in terms of whether and at what levels they can provide decision-making assistance, or whether their relevance is limited more to obtaining and displaying high levels of information for decision-making. Consequently, further examination of expert systems is beyond the scope of this book.

Decision Outcomes

The reader is referred back to Figure 1.1 which shows the limits placed on the examination of decision-making in OCCs to the point at which a decision is made to resolve some disruption. In the study, a panel of experts in each of the international and domestic simulations provided several decision outcomes as viable solutions to a series of operational problems. However, the stage of reaching and evaluating *specific* decision outcomes from which a course of action could be implemented was beyond the focus of the study. In other words the intent of the study was not to identify and rate solutions reached by controllers as they solved a number of operational problems. Rather, the intent was to explore the ways in which they arrived at this juncture. Consequently, the actual decision outcomes were not taken into account in the study and the focus was on the decision-making processes to the point of generating decision alternatives.

Summary

This chapter provided some of the background research that underpins decision-making styles. Of note was a discussion regarding the benefits and limitations of rational decision-making and also the relevance, especially in OCCs, of intuitive decision-making as this appears to be a dominant approach used by decision-makers in this type of environment. Examination of decision-making in a naturalistic environment has gained momentum as researchers have established the need to explore ways in which people decide courses of action in the field rather than in contrived and orchestrated situations. Finally, the need for further study in several areas has been mentioned as a means for exploring ways in which decision-making can be augmented.

Chapter 4
Situation Awareness

Introduction

Situation awareness is a concept that emanated from the civil and military aviation environment, having been created by pilots to describe the degree of adaptation of a pilot to the work environment (Flach and Rasmussen 2000). The research on situation awareness has emphasised its relevance for decision-making and suggests that an understanding of situation awareness may assist decision-makers to achieve better problem solutions. Of interest also is the way in which situation awareness may be acquired and the extent to which levels of experience are likely to contribute to this process. It appears that situation awareness of controllers in OCCs has not been examined in any detail, so this chapter provides some formative background to aid the reader's understanding of situation awareness.

Definition of Situation Awareness

Defining situation awareness has been very difficult for a number of reasons. One difficulty has been differentiating between situation awareness either as a *process* or as a *state*. For example, Sonnenwald, Maglaughlin and Whitton (2004: 991) define the term as '... a sense of knowing about things that are happening in the immediate environment and having both an understanding of the situation and the knowledge to respond appropriately as the situation evolves'. Sarter and Woods (1991: 52) regard it as '... the accessibility and coherent situation phenomenon which is continuously being updated in accordance with the results of recurrent situation assessments' and Klein (1993) refers to gaining situation awareness as a process of acquiring knowledge.

In contrast, Adams, Tenney and Pew (1995: 85) describe situation awareness as an up-to-the-minute state of '... cognizance required to operate or maintain a system'. Itoh and Inagaki (2004: 5) regard it as '... a snapshot of an operator's understanding of a situation' and Endsley (1988) describes it as a 'state' of knowledge, carefully distinguishing the state of knowledge from the processes used to derive it. The confusion in terminology is evident further in Matthews, Strater and Endsley's (2004) definition which suggests that situation awareness refers to the cognitive processes involved in perceiving and comprehending the meaning of a given environment. So an understanding of situation awareness appears to embrace both knowledge of a situation and the processes that produce the knowledge; an explanation which further indicates the difficulty in defining situation awareness clearly.

Another difficulty in defining situation awareness has been differentiating situation *awareness* from situation *assessment*. Klein (1993) refers to situation assessment as central to decision-making and extends the meaning of assessment to include the identification of goals, cues, and actions. However, this interpretation does not appear to be shared in the majority of the literature. Endsley (1995b) and Sarter and Woods (1991) clarify the distinction between the terms by explaining situation assessment as building knowledge that leads to the acquiring and maintaining of situation awareness. Sarter and Woods (1991: 50) regard situation awareness as being '... based on the integration of knowledge resulting from recurrent situation assessments'. In light of these explanations, situation *awareness* appears to be regarded as the state of knowledge and situation *assessment* as the process from which that state is derived. Regardless of these views, Salas, Prince, Baker and Shrestha (1995) note the dynamism of situation awareness and its need for continual modification and updating. To minimise confusion and provide a focus in this book, the term situation awareness is used in relation to understanding decision-making processes of controllers in OCCs.

Despite numerous attempts to define and a degree of confusion in the usage of terminology, no definition of situation awareness appears to be universally accepted. However, Endsley's (1988: 5) definition as '... the perception of the elements in the environment within a volume of time and space, the comprehension of their meaning, and the projection of their status in the near future' has been relied on extensively in recent literature and provides an appropriate definition for this book.

Importance of Situation Awareness

The importance of situation awareness has been explained in terms of being well-informed, achieving enhanced readiness for decision-making, being able to reduce operational error, and predicting operator competence in complex environments (Croft, Banbury, Butler and Berry 2004). Building on early research emanating from the aviation domain (e.g., O'Hare 1997, Robertson and Endsley 1995), the realisation of the importance of situation awareness has broadened research in other complex, dynamic environments such as anaesthesiology, military flying, marine navigation, driving performance, and medical emergency despatch. Situation awareness has been considered in terms of decision-makers acquiring a mental picture as a precursor to responding in a situation. For example, in their study of situation awareness of emergency medical despatchers, Blandford and Wong (2004: 448) noted that ambulance allocators (N=18) referred to '... the knowledge of the situation as a mental picture' and to the necessity to update the picture as circumstances change. Thus, it is evident that decision-makers in these domains recognise the necessity to accumulate and maintain situation awareness.

Situation Awareness in Aviation

In aviation, much of the research on situation awareness has related to piloting. This is because of the prominent role situation awareness plays in flight operations, in particular with regard to the safe conduct of the aircraft (Sohn and Doane 2004). According to Endsley (1988), the best trained and most experienced pilots can make wrong decisions if their situation awareness is incomplete or inaccurate. Therefore, they spend a lot of time building awareness and preparing for potential problems.

Situation awareness in aviation has also been examined extensively in relation to air traffic control (e.g., Hauss and Eyferth 2003, Niessen and Eyferth 2001). It has emerged from these and other studies that forming mental pictures is a key aspect for helping air traffic controllers develop and maintain situation awareness. This enables controllers to gain a clear 'picture' of aircraft positions and movements from which they can make informed decisions and take appropriate actions.

Situation Awareness in OCCs

An airline OCC environment is extremely complex and dynamic. To familiarise themselves with potential operational problems and remain alert and informed during current problems, it is likely that OCC controllers need a high degree of awareness as a foundation for decision-making, in a similar way to pilots and air traffic controllers. However, very few, if any, studies appear to have examined situation awareness in OCCs, so this is a major focus of the study discussed in Part Two. It is important to note that although most research on situation awareness has focused on individual levels of awareness, other research has considered *team* situation awareness in terms of aircrew member role assignment (Jentsch, Barnett, Bowers and Salas 1999) and low experience level pilots (Prince, Ellis, Brannick and Salas 2007). These studies have examined cooperative practices between team members and in particular, the ways in which team members accumulate levels of situation awareness, respond to changes in situations, and behave should they lose awareness. Examination of team situation awareness in OCCs was beyond the scope of this book, but future research is certainly warranted to explore the relevance of this to decision-making in OCCs.

Acquiring Situation Awareness

Many studies have indicated that experience is an important precursor for acquiring situation awareness (e.g., Blandford and Wong 2004, Kass, Cole and Stanny 2007). Some suggest that experience helps to reduce the mental resources needed for building and maintaining awareness (e.g., Sohn and Doane 2004), while others warn that a lack of experience may actually lead to a loss of situation awareness (e.g., Wiegmann, Goh and O'Hare 2002). This raises questions as to the extent of situation awareness gained by novices and experts implying that experts are likely to develop better situation awareness as a result of their superior levels of experience. There does not seem to be consensus in regard to gaining an optimal level of situation awareness; rather an appropriate level which may be suitable and may vary according to the nature of evolving situations. Decision-makers must be cautious of relying on already accumulated situation awareness to the exclusion of gaining further or updated awareness, as this is likely to result in insufficient or incomplete situation awareness which may lead to incorrect decision-making. So, the acquisition of situation awareness appears to be a cumulative process which may be achieved in a number of stages, but ascertaining what levels may be appropriate for decision-making in OCCs is of interest. A key focus in Part Two is exploring the extent to which controllers gain a level of initial awareness and whether they build on this to develop more comprehensive awareness in response to changing circumstances.

Information Completeness

Decision-makers need to gather relevant and sufficient information in order to ensure that decision-making is based on complete rather than partial advice. At the same time, decision-makers need to avoid saturation or overloading of information. The right quantity and importantly, quality of information is crucial for sound decision-making, where quality of information refers to its usefulness, currency, and accuracy. While attention has been given to the need for information to gain situation awareness, there are difficulties in achieving this due perhaps to the inability of many individuals to acquire the right information at the right time or their failure to ensure that they capture information from a variety of sources. In airline OCCs, gathering sufficient and appropriate information for decision-making is critical due to the complexity of problems encountered in a constantly changing environment. This complexity creates a high level of uncertainty which may be complicated further by conflicting and often technical information requiring expert interpretation. A further obstacle in OCCs may be sourcing information that is not readily forthcoming. For example, information relating to engineering related activity may be typically available for the OCC, but information from less critical functions like check-in control or gate-lounges may not be. Thus, it is important for controllers to acquire precise and complete information in a timely manner as this is likely to lead to informed decision-making. This is a challenge for decision-makers and is considered in Part Two by examining the extent to which controllers draw on information and then considering ways in which this information enables them to develop situation awareness.

Situation Awareness and Decision-making

Much of the discussion examining situation awareness emphasises the relationship between situation awareness and decision-making. According to Kaber, Perry, Segall, McClernon and Prinzel (2006), this emphasis is based on the need to develop situation awareness in an environment with multiple competing goals, complex decision requirements, and simultaneous demands with limited time for decision-making. Also advocated is a relationship between situation awareness and performance of decision-makers, although the extent of this relationship seems arguable. For example, high levels of situation awareness may not necessarily lead to high performance and high performance may still be evident with minimal situation awareness. Other arguments suggest that participants with advanced situation awareness are quite likely to perform at a superior level (e.g., O'Brien and O'Hare 2007). Klein (1998) supports this latter view and suggests that performance is dependent on achieving good situation awareness for recognition-primed decision-making. This relationship is evident in his RPD model depicted in Chapter Two. Despite these varying perspectives, it appears that situation awareness is at least one major determinant of superior decision-making.

While there is evidence of a relationship between situation awareness and decision-making, it is also evident that situation awareness is not part of the decision-making process. Gaining situation awareness may enable effective decisions to be made. However, there is no certainty as to which, if any, decisions need to be made having gained this awareness. Nevertheless, it appears that achieving situation awareness in a constantly evolving environment is regarded as a pre-requisite for good decision-making. Figure 4.1 presents a grossly simplified adaptation of Endsley's model of Situation Awareness. The source for the complete model is Endsley (1995a) in the reference list at the end of this book.

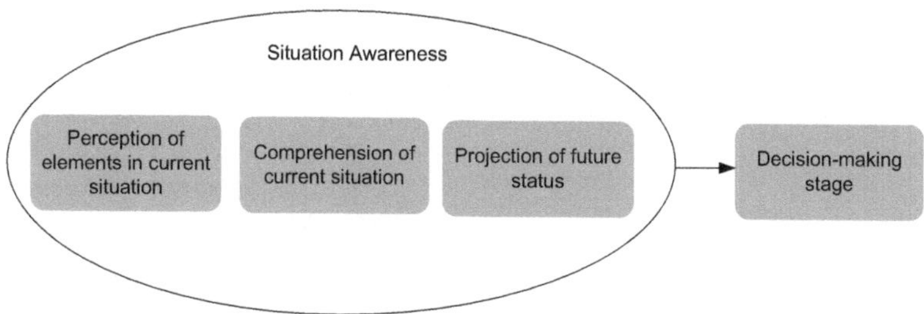

Figure 4.1 Simplified adaptation of Endsley's model of Situation Awareness

The model clearly depicts the separation between situation awareness and decision-making. In terms of controllers in OCCs, this should enable a better understanding of ways in which they are able to build knowledge by gaining awareness of a situation. Second, it should help to explain the influence of situation awareness on their decision-making processes. In light of Endsley's model, this draws attention to the perception stage (level one) of situation awareness. Notably, Endsley's model is not totally inflexible as it can be updated from feedback, and inputs from the environment, which means that it can provide a mechanism to examine ways in which OCC controllers accumulate initial situation awareness and importantly further situation awareness based on changing circumstances. In light of the nature of operational disruptions, the model seems an appropriate vehicle for examining situation awareness in OCCs.

Summary

Plenty of research has examined situation awareness across several industries, but little work has been done in OCCs. This chapter has emphasised the importance of situation awareness and its relevance for decision-making in OCCs. In addition, comment was provided in terms of the need for sufficient information to acquire and maintain situation awareness.

Chapter 5
Expertise

Introduction

Considerable research has been conducted over a long time with regard to the influence of expertise on decision-making. Quite a bit of this work has come about as a result of wishing to know what experts do and how they do it and the extent to which learning from experts may lead to improvements in decision-making by novices. Very little work to date appears to have examined the influence of expertise on decision-making processes of OCC controllers. Consequently, little is known in terms of determining any guidance that could be provided to novices to improve their decision-making processes in OCCs. Thus, a focus of the study in Part Two is to examine decision-making processes of controllers according to their levels of both industry and OCC expertise.

Definition of Expertise

Defining expertise has been particularly difficult because of the vast number of approaches taken. For example, Johnson, Zualkernan and Tukey (1993) propose that types of expertise correspond to the types of problems an expert may solve. Then, knowing the type of expertise may help to explain the actions taken by a decision-maker and determine the basis for success or failure of the decision. Other approaches have included categorisation of expertise by a consensus of agreement, acknowledgement of respondents' own reported levels of confidence in their experience, and expertise expressed in terms of observed performance. Expertise has been referred to as domain specific to the extent that expertise requires *knowledge* in that particular domain, knowledge of *problems* in that domain, and skills to *solve* those problems (Hayes-Roth, Waterman and Lenat 1983). There is general agreement, though, that knowledge and skills are regarded as fundamental for gaining mastery in a domain (Hedlund and Sternberg 2000).

In a substantial review of the previous work on expertise, Shanteau, Weiss, Thomas and Pounds (2002) critiqued nine traditional approaches for identifying experts. These approaches include: classifying experienced people as experts, bestowing the title of expert, attributing an individual as an expert, self-acclaim of expertise, identifying expertise by consistency in judgments, reaching consensus between experts, having an ability to discriminate, displaying knowledge, and creating expertise as a result of specific training. Although Shanteau et al. (2002) identified applications and benefits of these approaches they also determined that each approach had flaws, concluding that none of the definitions of expertise was generally acceptable. Despite these varied approaches and the criticisms, there is abundant evidence in numerous studies to date suggesting that expertise draws heavily on experience (e.g., Chi, Glaser and Farr 1988, Sonenshein 2007). In fact, it appears that experience may well be the most important factor in the acquisition of expertise.

There also appears to be ample agreement that ten years' experience in a specific domain is required for gaining expertise in that domain (e.g., Hershey and Walsh 2000, Simon and Chase 1973). Despite some views that experience may not be the best predictor of expertise, the common use of experience in much of the literature on expertise is sufficient reason for classifying experts according to experience. In contrast, novices have been described as having exposure to a domain, but to a far limited extent (Hoffman, Shadbolt, Burton and Klein 1995). Taking these perspectives into account, the study in Part Two relied on ten years' experience as a basis for defining expertise at both industry and OCC levels and in particular, as a means of distinguishing between novices and experts.

Importance of Expertise

Expertise has been studied over several decades in numerous fields. The studies have been important as they have identified ways in which experts categorise problems, gather and use information, and use their knowledge to devise strategies and solve problems. However, it is evident that one of the most important goals for studying expertise is '... as a means of improving the training of less skilled individuals' (Ericsson 2005a: 238). In terms of the study discussed in Part Two, expertise of controllers is likely to be very important for airline OCCs due to the need for rapid and accurate decision-making in such a dynamic, challenging environment. Thus, a key focus of the study is to examine differences in decision-making processes by novice and expert controllers with a view to identifying ways in which novice decision-making processes can be improved.

A domain of great interest has been the study of decision-making processes of players during chess games due to the complex tasks required to play the game and the extensive training to reach a professional level. De Groot (1978) analysed thought processes as they occurred in chess by conducting short-term memory experiments with chess players (N=22) at grand-masters level and below and found that chess masters recognised patterns among chess pieces. Chase and Simon (1973) suggest that it is not the memories of chess players, nor the depth of planning moves ahead that set experts apart from others. Rather, experts produced superior performance through their ability to look at large, complex displays and see them in terms of meaningful chunks. Airline OCCs use complex and colourful displays to represent aircraft flight schedules (see appendices A and B and note the differences in complexity between the international and domestic schedules). This early research on expertise provides a foundation for examining ways in which novices and experts in airline OCCs derive relevant and meaningful information and the extent to which they identify patterns within the flight displays.

Research has focused extensively on domains calling for decision-making in critical circumstances such as pilot communication and air traffic control. A common emphasis in this work has been the need for rapid responses in dynamically changing situations and the need for expertise of decision-makers to ensure that safety and security, as well as operational and commercial efficiencies are met. This has led to studies of expert behaviours that focus on superior performance, limitations to performance, and strategies to overcome these limitations (e.g., Gobet and Simon 2000).

The conclusions reached from research on expertise have drawn attention to the ways in which experts endeavour to understand problems and how they complete tasks fast, effectively, and efficiently in order to produce high quality solutions. However, there appears to be an assumption evident from this work that studying ways in which experts make decisions will somehow enable a clear understanding of decision-making by novices. This may not necessarily be the case. Examining decision-making processes of both novices and experts in OCCs should help to determine the extent to which expertise influences decision-making processes in this domain and identify ways to improve decision-making of novices.

Characteristics of Experts

Experts display many characteristics that set them apart from novices or those not yet considered to be expert. For example, experts appear to excel in their own domain of expertise, perform largely free of errors, and exhibit deep problem representation. Experts also depend on a highly organised base of relevant knowledge from which they are able to assess situations quickly and correctly. This knowledge is accumulated by identifying information or sensing cues from a number of multi-dimensional stimuli and doing this by using small quantities of information. What is crucial is experts' ability to discriminate between information that is *task relevant* and *task irrelevant* (Haider and Frensch 1999). A number of studies have considered how experts are able to do this. For example, Selnes and Troye (1989) suggest that experts frame situations to ascertain their information requirements. That is, they work out what has to be done in a specific purpose or event and then acquire and limit the information to help them achieve their task. More recently, Carr and Wagner (2002) suggest that experts use their knowledge to classify tasks into a series of categories from which they can prioritise their work.

Other research has recognised that experts are able to explain a problem situation by means of collecting and evaluating critical information (e.g., van den Bosch and Helsdingen 2002). This enables them to comprehend and assess situations quickly and anticipate future events, resulting in outstanding problem solving performance. According to Cellier, Eyrolle and Marine (1997), the ability for experts to anticipate events is underpinned by the recognition of cues during monitoring and diagnosis of situations. This is because differences in the ways in which novices and experts recognise cues are likely to result in differences in decision outcomes. In terms of airline OCCs, controllers need to identify and analyse problems in great depth before reaching any decision points. Therefore, it would seem most likely that assisting controllers to recognise appropriate cues would help them anticipate situations more effectively which in turn should benefit decision-making in the OCC.

Acquiring Expertise

The acquisition of individual expertise emanates from numerous sources. Some arguments associate expertise with innate basic capabilities, unable to be influenced by training or practice while others strongly emphasise the role of experience in gaining expertise and more-so the importance of variety and meaningfulness of experience to build expertise (e.g., Jentsch, Bowers and Salas 2001). This is achieved through exposure to novel situations and multiple types of experience rather than by repetition of similar events. Several researchers contend that acquiring expertise is subject to a combination of experience and domain knowledge (e.g., Hershey and Walsh 2000, Weiss and Shanteau 2003). It is also evident in past studies that expertise is gained through deliberate practice and intense preparation (e.g., Ericsson 2005b, Ericsson and Lehmann 1996). The work of Bedard (1991) and Hershey and Walsh (2000) suggest that novices who undergo task-related training in the domain are likely to produce better solutions than experts. This is an interesting observation and anecdotal evidence can also attest to the ability of *non-experts* being able to solve quite complex operational problems with little experience and even any formal training. Thus, it appears that expertise alone may not account necessarily for superior decision-making.

Differences Between Novices and Experts

Examining difference between novices and experts has long been an area of great interest. According to Cannon-Bowers, Salas and Pruitt (1996), this focus has occurred for a number of reasons. First, it may not be possible to determine who is expert and who is not. Second, important decisions are made by people whose expertise is questionable or who may not be experts and third, decisions made by novices may be significant for developing decision strategies. Therefore, identifying differences between novices and experts is likely to provide a basis for improving decision-making. This is relevant in particular for domains such as airline OCCs where decision-making is critical.

The research has established that novices and experts tend to make decisions and solve problems quite differently. According to Orasanu and Connolly (1993:

12), experts '... use their knowledge and experience to size up a situation, determine if a problem exists, and decide whether and how to act on it'. There is ample evidence that experts are able to recognise and discern between surface and underlying features of a problem. That is, they can determine the problem's level of complexity and then identify what can be dealt with quickly and easily and what will need further examination. This distinction enables them to draw all the components together to form an integrated picture of a situation. In comparison, novices only see bits and pieces and are more likely to deal in terms of the surface features of problems. So they may not have the experience or exposure to delve deeply enough in order to realise that a more complex problem even exists. According to Chase and Simon (1973) experts have been differentiated from novices also by the manner in which they organise the components of problems. Experts do this readily by concentrating on the nature of the problem and identifying the inter-relationship among components, enabling them to recognise correct solutions. Novices, on the other hand need to spend considerable time analysing the parts very carefully in order and don't necessarily piece them together to solve the overall problem.

Considerable interest has been shown concerning the progress of individuals from novice to expert (e.g., Helton 2004, Russo 2006). Dreyfus and Dreyfus (1986) contend that individuals go through a number of stages between novice and expert as they gain experience, developing various competencies and proficiencies before expertise is achieved. Thus, a continuum of development arises, namely: novice, advanced beginner, competent, proficient, and expert. Stages have also been categorised as advanced beginner, intermediate, and expert. However, most of the studies on expertise refer to a dichotomous classification: novice and expert, and this approach is taken in this book.

Expertise and Decision-making Styles

There is considerable agreement in the expertise literature suggesting that as practitioners gain experience and expertise, they use intuition in order to make decisions without having to rely on analytical thought to guide their decisions (e.g., Bryer 2006, Cesna and Mosier 2005). This is because experts rely on their knowledge of what normally works rather than a need to compare and evaluate decision alternatives. However, according to Benner (1984), experts may resort to analytical tools when they do not have experience in particular situations, incorrectly assess a situation, or when behaviours do not occur as expected. As noted earlier, the complex, challenging environment of an airline OCC provides situations calling for rapid and accurate decision-making processes. Taking into account the review of decision-making styles in Chapter Three, a focus of the book is to examine the decision-making styles of both novice and expert controllers in airline OCCs with a view to determining if either a rational or intuitive approach is more likely to lead to improved decision-making outcomes.

Summary

Another important component of decision-making is the level of expertise of the decision-maker. This chapter has reviewed some of the ways in which expertise has been defined and provided some theoretical background as to the importance of expertise especially in complex, challenging environments. The characteristics of experts were described as were differences between novices and experts. Finally, it was suggested that examining the different decision-making styles employed by novices and experts may raise interesting questions as to potential improvements in decision outcomes. The main theoretical frameworks upon which the study is based have been described in these early chapters. Without an over-emphasis on the literary aspects of this foundation, hopefully the reader will appreciate the rationale for the study. Part Two of the book describes in some detail the study conducted with 52 controllers, specifically applying these theoretical frameworks to the real workplace. In the case of airline OCCs, this form of application has been missing from previous studies so the intention of this study was to address this deficiency. Of course the study could always be more rigorous, more time consuming, more wide-ranging and covering a greater diversity of factors influencing decision-making and indeed, this is the intent of further work and research in OCCs. This study is a starting point and the reader should find the examples of protocols captured during the simulations insightful for gaining a small degree of understanding of decision-making in OCCs.

PART II
Examining Decision-making Processes of OCC Controllers

Chapter 6
Designing the Study

Introduction

This chapter examines the research design and methods used in the study. It also explains the purpose and design of the preliminary study, and the subsequent re-design of the study using simulations as a method to replicate real-life OCC disruptions. The use of simulations enabled the capture of thought processes of participants and provided a means for collecting quantitative and qualitative data. Co-incidental with this approach was the expansion of the study to a multiple case approach enabling comparisons to be made between decision-making processes of controllers in international and domestic environments. Finally, the chapter explains the methods for analysing the data.

Delimitations of the Study

As always with research of this nature, a number of limits were placed on the study due to geographical boundaries, time, optimal variety of OCCs examined and the need to satisfy objectives:

1. The study was confined to OCCs within the Asia-Pacific region;
2. The study clearly distinguished between decisions made with regard to safety as distinct from decisions made for operational reasons. For instance, safety-related decisions may be made by pilots or air traffic controllers where the wellbeing of an aircraft, persons, or the aviation system is potentially or actually compromised. However, while the *outcomes* of such decisions may be taken into account in airline operations, OCC decisions are not directly safety-related. Rather, they are operationally inspired. As an example, if a decision by an OCC required a specific direction to the Captain of an aircraft (e.g., to divert the aircraft), the decision will not compromise the responsibility that the captain has for any safety matters pertaining to that aircraft;
3. In some OCCs (notably those located in North America), licensed flight despatchers share responsibility for the issue and ongoing safety of flights, planning and monitoring the progress of flights, and taking actions in the event of disruptions. In Europe and Asia-Pacific regions, flight planning and despatch roles are performed outside the OCC. OCC controllers in these regions manage the aircraft resources during disruptions and liaise

with crewing, maintenance, and other support areas (Kohl, Larsen, Larsen, Ross and Tiourine 2007). Thus, the focus of the study was limited to the examination of decision-making processes of controllers in the Asia-Pacific region and excluded the roles of flight despatchers;

4. The study did not analyse the results according to particular airlines but comparisons were limited according to whether controllers were employed in international or domestic airlines;
5. While numerous aspects of personality may influence the decision-making process, these aspects were excluded from the study;
6. While it is recognised that risk analysis may influence the decision-making process, this aspect was excluded from the study;
7. While the study took into consideration the influence of time pressure on decision-making, levels of arousal and psychological stress on participants were considered to be beyond the scope of the study;
8. The study limited the number of simulation scenarios to three as a necessary constraint on time resources, controller availability, and manageability of the research;
9. The focus of this study was to examine the decision-making processes to the point of generating decision alternatives. Consequently, identification, selection and evaluation of the final (optimal?) decision and the resulting actions that could be taken by participants were not considered;
10. The study examined decision-making processes of individuals in OCCs. While the importance of team decision-making is recognised, the examination of teams was not considered in the study. This means that team situation awareness also was not examined;
11. While the study recognised that novice controllers in OCCs were likely to have received some form of training, the influence of training on decision-making processes of novices was not examined in the study;
12. The study did not consider knowledge management issues in regard to the transfer of knowledge between personnel in OCCs;
13. The study did not consider the influence of organisational structure on decision-making process; and,
14. The study did not consider the use of expert systems, decision support systems or other analytical tools to assist decision-making.

Assumptions

A number of assumptions concerning the research, sample and collection of data have been made:

1. Participants were representative of operations controllers in airline OCCs;
2. The simulations were representative of typical operational problems;
3. Participants were able to understand the requirements of the research;

4. In countries where the English language was not the primary language, participants had sufficient understanding of the language to participate fully in the study;
5. Participants answered all questions honestly and to the best of their ability;
6. Responses were an accurate reflection of participants' current decision-making processes; and,
7. The audio tapes were transcribed accurately and were true records of interviews conducted.

Data Collection Methods

When trying to examine phenomena, investigators often face a difficult task in attempting to capture data accurately in a way that is sound methodologically and so that generalisations can be made across a broader population. Common primary data collection methods include surveys (including questionnaires and interviews), observation, and experiments/simulations and all of these methods may have advantages and limitations subject to their suitability for specific research projects. For example, surveys are cost effective and easily distributed, but there may be difficulty as to the respondents' understanding of the questions or motivation to complete the required answers, whether events may be recalled in sufficient detail, and whether information can be verified and relied upon. In addition, they are not ideally suited for in-depth examination of the thoughts and feelings of individual participants as they can inhibit free-flowing speech, reduce spontaneity and constrain responses.

Observation enables researchers to gather data at the time behaviour occurs, open possible dialogue and other interaction with the participant and perhaps gain a clearer picture of responses than could be gained otherwise. But observer obtrusiveness may be problematic in terms of a protracted and/or costly process or there may be a lack of predictability of events. Worse still is that so many events may occur that observations of specific criteria become indiscernible or even impossible. Further, it may be very difficult to capture attitudes and thought processes or prevent observer subjectivity.

Simulations can mimic reality quite well and major variables can be manipulated to replicate specific situations so that researchers can study phenomena too difficult or impossible to study in real life. A good example of this is the flight simulator of course. But simulation requires a thorough knowledge about the simulated environment and the results may not necessarily reflect behaviour that occurs in the more complex operational environment. In conjunction with simulations, a technique called 'think-aloud' may be used, giving researchers access to participants' underlying thought processes, reasoning, and behaviours involved in analysing and solving problems, thus providing the researchers with a source of rich information. Analysis of these verbal protocols then enables researchers to examine the mental processes that may be revealed. Sometimes though, participants may not be able to put their thoughts into words, or express their mental processes sufficiently, especially when considering many dimensions simultaneously. In these cases, prudent encouragement and non-judgmental feedback from the researcher may lead to increased accuracy of responses. The main objective when examining decision-making in OCCs was to capture the thought processes of controllers as they attempted to manage disruptions. Therefore, selecting an appropriate method for the data collection method was very important.

Preliminary Study

A preliminary study was conducted in one OCC prior to the main study in order to explore themes initially identified by OCC management. A survey questionnaire was designed for participants to provide information about decisions they made, as disruptions occurred. However, this method was considered by OCC management to be disruptive to the work practices of controllers and prevented adequate collection of data. Another issue was the reluctance and often inability of controllers to engage in discussion about their thought processes during a disruption as they needed to focus intently on solving operational problems. Therefore, a questionnaire called a 'post-disruption feedback form' (PDFF) was designed as a means for collecting appropriate decision-making information following each disruption. The PDFF was completed anonymously, requiring participants to advise their position title, their duration of employment (years) both in the aviation industry and in the OCC, and their education level.

At the completion of each disruption, participants were requested to provide information about the nature and cause of the disruption, the extent to which their previous experience influenced their decision-making processes, and the extent of their situation awareness during the disruption. Participants were required to complete a new PDFF for each disruption. The cause of a disruption could be chosen from several options such as 'weather', 'engineering', 'air traffic control', and 'ground handling'. OCC management validated these options by providing feedback during the design stage of the questionnaire. Other questions were designed to investigate information sought by participants such as: 'where did you obtain information for decision-making?', 'what amount of information did you need to make decisions?', and 'how much was your decision based upon recognition of a similar problem?' A comments box was provided for participants to add general comments in relation to the disruption or to their own role in decision-making.

Limitations

Several factors were beyond the control of the researcher in the administration of the PDFF. First, the researcher had little influence over the distribution of the survey. Although OCC management assured the researcher that participants would be requested to complete the PDFF, there was no control mechanism to ensure this occurred. Second, there was no way of determining who had actually completed the PDFF and when it had been completed. During the data collection stage of four weeks in this particular OCC, 15 disruptions had been recorded and some of those had continued across a number of shifts or even days. Therefore, decision-making processes of any one participant and for any one disruption were difficult to capture. Third, there was no way for comparisons of participants' decision-making processes. Some disruptions were resolved quickly requiring limited decision-making from controllers, while others were longer requiring extensive decision-making. Fourth, if a number of disruptions occurred simultaneously, decision-making relevant to any particular disruption could not necessarily be determined. Fifth, there were no means by which the researcher could be confident that participants completed the questionnaire with any degree of accuracy, honesty, or timeliness.

Outcome and Implications of the Preliminary Study

The conclusion reached from the preliminary study was that the method of using a post-disruption survey to study decision-making in an OCC was inappropriate. Further, the study was predicated on self-reflection by controllers and accurate analysis of decisions possibly well after decision-making occurred. This raised questions as to whether the data may be comprised due to accurate recall of events, possible errors and bias. As a consequence of these limitations and the anticipated difficulty in achieving reliable data, the preliminary study was abandoned and the study re-designed.

Main Study

The study was expanded to investigate decision-making processes in six OCCs selected from the Asia-Pacific region. This multiple case approach helped to increase the volume of data, provide more rigorous results and deepen the understanding of operational decision-making in such a complex area. Access to each of these OCCs was readily forthcoming and staff and resources were made available for the timely and efficient collection of data. No OCC management indicated concern for the findings of the study.

By selecting OCCs of airlines operating international schedules and domestic schedules, decision-making processes in each environment could be examined. Simulations were constructed to facilitate this approach ensuring that control could be exercised over the conduct of disruptions and responses of participants could be identified and appropriately recorded. In order to examine decision-making processes in international and domestic environments, the study involved OCCs in Australian domestic airlines, as well as OCCs of international airlines in the Asia-

Pacific region. One airline operated solely international services, two operated solely domestic services and two operated both international and domestic services. Advice was sought from an expert panel of current and/or retired senior OCC management staff to ensure that proposed OCCs were representative of airline OCCs around the world, and to ensure important and relevant issues were to be investigated.

Generally, in international operations flight stages may be very long (up to 18 hours), aircraft may only operate a small number of flights per day, and are on the ground between flights for long durations (often several hours). In general terms, decision-making in international operations occurs over long time-frames. In contrast, flight stages in domestic operations typically are short (often between one and four hours), aircraft are scheduled to operate a high number of flight stages in a day (10–12 stages) and usually remain on the ground for short durations between flights (usually between thirty minutes and one hour). Controllers handling domestic disruptions make numerous decisions within very short time-frames. Consequently, decision-making in domestic operations is far more intense than decision-making in international operations. These differences provided the impetus for examining decision-making processes in each of the international and domestic environments.

Ethical Considerations

Prior to conducting research such as this study, it is imperative to respect and protect the participants' welfare and any information they provide. This means that the research has to ensure that participation is voluntary, anonymous, and confidential, and that no harm may come to participants. Further, the researcher has an obligation to protect the anonymity of participants. As a result, proper permissions were received prior to any data collection.

Explanatory Statement and Consent

A written explanatory statement provided to airline management and each participant prior to conducting the research outlined the nature and procedure of the research, emphasised the anonymity of participants and their company, and guaranteed confidentiality of the results. Participants were advised that they would be interviewed and all communications would be audio-taped. Further, participants were advised that they could access the audiotape of their interview on request, although no participants ever asked for this access. Airline management were also able to obtain aggregated results. Participants were advised that they could withdraw at any time, but no one withdrew from the study. Finally, a consent form was provided to airline management and each participant prior to conducting the research to ensure that participants agreed to participate in the study and agreed that their comments could be audio-taped.

Collecting Data in the Workplace

It is very useful to examine attitudes and behaviours of subjects within their natural settings rather than in artificial settings of experiments and surveys but sometimes it is very difficult to gain access to the workplace. This is often the case if the researcher has yet to establish credibility with either management or participants. In terms of this study, significant previous employment of the researcher as a controller in an OCC provided a high level of familiarity and personal experience. Thus, rapport, mutual respect and importantly, trust were established early. This was essential to enable the researcher to explore the controllers' decision-making processes in considerable depth. To facilitate the process, management staff in the OCCs briefed participants with respect to the purpose of the study and the background of the researcher.

Designing Simulations

Simulations were designed in collaboration with two independent experts. One expert who had extensive experience in international OCCs assisted with the design of the international simulations. Another expert with extensive experience in domestic OCCs assisted with the design of the domestic simulations. The simulations were shown to four other experts for feedback and suggestions for any changes. The value of expert input into the design of the simulations was to ensure the simulations were representative of realistic and sufficiently complex disruptions such that the decision-making processes could be thoroughly investigated. None of the experts participated in the study. All information and communication during the data collection stage was conducted in English which is the international language for aviation.

International Simulations

One set of simulations typified a sample of flight schedules for an international airline operation. The airline base for these simulations was created as a fictitious locality in the Asia Pacific region and was given the name 'Pacific'. Due to the involvement of OCC controllers from several Asia-Pacific airlines, the siting and naming of this base avoided bias or inequity in relation to any particular airline OCC involved in the study. Information for participants included the approximate location and local time zone of Pacific as depicted on a map supplied at the time of the simulation. Participants could determine the flying time and therefore distance from Pacific to any destination by consulting the flight schedules provided. In the international simulations, a fleet of 18 aircraft comprised six large (long-haul) aircraft capable of carrying 450 passengers and 12 smaller (short-haul) aircraft capable of carrying 250 passengers. The members of the expert panel with international OCC decision-making experience agreed that the flight schedules and aircraft sizes and types selected were appropriate for the international simulations.

Domestic Simulations

The other set of simulations typified a sample of flight schedules for a fictitious domestic airline operation. This airline was based in Australia and controllers who participated in the domestic simulations were employed in airlines based either in Australia or New Zealand. Therefore, these controllers were familiar with airline schedules in Australia and the airports to which Australian airlines operated. In the domestic simulations, a fleet of 20 aircraft comprised six large (wide-body) aircraft capable of carrying 250 passengers and 14 smaller (narrow-body) aircraft capable of carrying 150 passengers. The members of the expert panel with domestic OCC decision-making experience agreed that the flight schedules and aircraft sizes and types selected were appropriate for the domestic simulations.

Flight Displays

The flight schedules were presented on a static, visual display showing a coloured utilisation of the fleet of aircraft for one complete day's flight schedules and part of the following day's flight schedules. The visual displays were appropriate for the study as they were easily transportable between OCCs, did not need compatibility with other airline infrastructure (e.g., computer equipment), and enabled participants to view an entire operation at one time without having to move parts of the display. Each display was laminated to enable participants to write or draw on its surface. On the horizontal axis, across the top and the bottom of each display, was a time-scale showing hourly gradations. On the left hand side of each display, a vertical scale showed fictitious aircraft registrations to identify each aircraft in the fleet. Types of aircraft were colour coded for easy identification. A secondary vertical scale showed the number of hours an aircraft could be flown before requiring a maintenance service. For example, the time 12.15 beside one aircraft registration indicated that the aircraft could be flown a maximum of 12 hours and 15 minutes before requiring a maintenance service. On the right-hand side of the display, the aircraft registrations were repeated and any maintenance requirements such as planned maintenance services and engine changes were also shown.

The flight schedules designed for the simulations were constructed to replicate typical airline schedules. Each flight was represented by a block of information (flight block or sometimes called a puk) comprising a flight number (white on a black background), actual passenger loading for the flight (yellow), departure airport and time (white), and arrival airport and time (white). The background colour of each flight block was designed to represent a type of aircraft as described above. Controllers were able to identify the duration of flights by the relative size of the flight block. For example, an eight hours and twenty minutes flight from Pacific to Melbourne was shown as one flight block stretching over that amount of time along the time-scale. If flights operated beyond the time-scale shown in the display, the flight block was truncated by a series of black crosses. For example, the international simulation showed long-haul flights which commenced on one day and arrived on the next day. This discontinuity was a limitation of the visual display and an explanation of this limitation was given to participants in the briefing tape. The flight blocks were positioned on the display according to the continuous time scale used in the display. The visual display was a close representation of the computerised display normally used by operations controllers.

Display for International Schedules

For the international simulations, the time-scale was shown in Coordinated Universal Time (UTC), commencing at 2100 and finishing at 1500. This scale was formerly known as Greenwich Mean Time (GMT). International OCCs use UTC time. Therefore, controllers were familiar with the time-scale used in the international simulations. A secondary, horizontal time-scale showed local time in Pacific. The local time zone for Pacific was selected as UTC plus nine hours. The range of the time-scale was selected to replicate the most common times during which short-haul international flights operate. As long-haul flights operate throughout a 24-hour period, the design of the display did not rely on these schedules. Appendix A presents a replication of the flight schedules for the international simulations.

Display for Domestic Schedules

For the domestic simulations, the time-scale was shown in Australian Eastern time for the day's operation, commencing at 0600 and finishing at 0100 the next day. The range of the time-scale was selected to replicate the most common times during which domestic flights operate. At the time of data collection, OCCs in Australia operated according to a similar time-scale. Therefore, controllers were familiar with the time-scale used in the domestic simulations. Vertical, dotted lines representing hourly divisions enabled the flights to be viewed on the displays according to scheduled departure and arrival times. A movable string time-line could be positioned along the time-scale axis by the researcher to indicate the specific time of day. For example, the time-line could be positioned to indicate the time at which a scenario commenced, or be re-positioned to indicate that a passage of time had elapsed during a scenario. Appendix B presents a replication of the flight schedules for the domestic simulations.

Simulation Scenarios

A series of simulation scenarios was designed in collaboration with the panel of experts. The scenarios consisted of three typical operational problems for each of the international and domestic operations. The three scenarios for each of the international and domestic simulations were completely independent of each other. Each successive scenario was more complex than the previous one, in terms of the nature of the disruption, the time-frame for decision-making and the consequences of decisions. A short audio briefing tape was made for each of the international and domestic simulations to ensure that all participants received a consistent briefing prior to the start of the scenarios. The tape provided participants with the rationale for the study and information about the background and operational experience of the researcher. The tape also provided a detailed description of the static display used in the simulations and a description of the process by which the simulations were to be conducted. In particular, the tape outlined the roles that were to be played by the researcher during the simulations. One role was to conduct the simulations and provide scenario or other information for participants according to a script. The other role was to act as a resource base and simulate the communications with airline departments such as crewing, maintenance, and airport handling as well as other aviation facilities such as air traffic control with whom controllers typically liaise during disruption management. This role helped to enhance the realism of the simulations.

The International Simulation Scenarios

In the international simulations, Scenario One consisted of a passenger connection problem. Participants were provided with initial information and asked to decide whether they would delay a long-range international flight in order to wait for 55 inbound passengers arriving from another flight. Scenario Two consisted of a mechanical problem with an aircraft on the ground at a European airport. Participants were asked to decide what action they would take as a result of the initial information received and what action they would take once the mechanical problem had been resolved. Scenario Three consisted of a major weather problem (approaching typhoon) at the main airport of the airline. Participants were asked to indicate the actions they would take in preparation for the weather disruption.

The Domestic Simulation Scenarios

In the domestic simulations, Scenario One consisted of a passenger connection problem. Participants were asked to decide whether they would delay a domestic flight in order to wait for 25 inbound passengers arriving from a delayed international flight. Scenario Two consisted of a mechanical problem with an aircraft in flight. Participants were asked to indicate what actions they would take in relation to an aircraft that had experienced a mechanical problem. Scenario Three consisted of two concurrent problems. Participants were asked to make several decisions about flights as weather conditions (fog) deteriorated at two airports. At the same time, participants were asked to make an aircraft available from the utilisation to operate two commercial charter flights.

Commencing the Study

Each simulation was conducted between the researcher and one participant at a time at the participant's workplace, but in a separate room away from the operations room. Participants were invited by their managers to participate voluntarily in the study. As only one OCC controller from any airline declined to participate in the study, the sample was highly representative of the population of controllers in the airlines studied. No participants withdrew at any stage during the simulations. In the OCCs of the two airlines in the study that operated both international and domestic operations, participants worked either as international or domestic controllers. These participants were permitted to select either the international or domestic simulations.

Prior to running the simulations, each participant listened to the five-minute audio briefing tape. At the conclusion of the briefing tape, participants were asked if any information required clarification or whether they needed additional information. When participants indicated their readiness to commence the simulation, the tape was changed and the tape recorder was set to record all subsequent communications, providing an accurate description of the critical events in the research setting. Taping the simulations provided two important benefits. First, original thoughts and words of participants were preserved and second, participants were assured that the record of what had been said was accurate. During the study, the taping process did not appear to present any interruptions to the simulations and the participants did not react adversely to the presence of the tape recorder, even to the extent of withholding comments while tapes were changed.

The researcher commenced the simulations by showing participants the utilisation of flight schedules and describing the display. The participants were asked to familiarise themselves with the utilisation and communicate observations and any questions to the researcher. As each scenario was introduced, participants were asked to think aloud as they considered the operational disruptions. During each scenario, real-time information was given to participants. On request from participants, the researcher provided information or answered questions relating, for example, to crew availability or roster commitments, passenger connections, airport or air traffic control related information, maintenance requirements, and weather conditions.

As the scenario disruptions were introduced, participants attempted to construct solutions to the problems. By observing participants' actions and listening to their verbalised thought processes, the researcher could gain insight into participants' decision-making processes, and explore specific themes further. So, not only was information collected about 'what' the decision-makers were thinking about, but importantly 'how' they were applying this approach. For example, if a participant commented about a flight with few passengers (proportion of booked passengers, to available seats), the researcher could probe the reason for the participant's comment. The researcher also asked questions such as 'why do you doubt the information provided by Engineering?' or 'how do you know when you have enough information to help you make a decision?'

An important dimension in observation is the ability of the researcher to identify acts such as physical movements, facial expressions or motor skills of participants. During the simulations, participants often became silent, but made physical movements such as pointing, so tolerance by the researcher for these silent periods enables observations that would otherwise have been missed. At these times, prompts were made to address the problem of un-surfaced thoughts. These prompts were kept short and nondirective to minimise awareness of the researcher's presence or interference with participants' thought processes.

Sometimes, a difficulty with the think-aloud method of data collection may be the occurrence of unreported information. So, if a participant did not verbalise thoughts, or made abrupt responses, probing questions such as 'what are you looking for now?' or 'what are you thinking?' were asked. This helped elicit more information about an earlier response or clarify comments. Exercising some caution to guard against biasing participants' thoughts or actions, these prompts actually contributed toward a conversation between the researcher and participant. In fact, in some cases, participants actively role-played the simulations. For example:

> 'Mr Crew Man, how far can this crew be extended?' and, 'OK, Mr Met Man, could you just give me a heads up on the current weather at Pacific?' (Case number: 307[1])

Another participant (308) conversed freely with the researcher:

> (Participant) 'This, this problem we've got with [aircraft registration] Hotel Bravo.'
>
> (Researcher) 'Yeah, well we've just heard about it, possible pump change. We don't think we've got a pump in CDG [Paris] but we're checking at the moment. We'll let you know when we've ... as soon as we hear.'
>
> (Participant) 'OK, when do you think you'll know by?'

---

1 A unique case number according to a specific airline OCC was assigned to each participant.

(Researcher) 'Well, we're on the phone now so it won't be too long.'

(Participant) 'OK. And is there any test run required after the pump change or is it two hours to complete that?'

It appears from the conversational approach used by participants that the role plays were indicative of the way in which participants may respond in real-life disruptions in OCCs. These responses also suggest that the simulations were representative of controllers' workplaces in OCCs. While no time limits were placed on the simulations, the scenarios were allowed to continue until the participant had either generated an alternative or the researcher determined that no further information would be gained by extending the time allowed. As a result, the simulations with each participant were conducted typically for one hour.

Data Transcription

Once the tapes had been transcribed verbatim, it was evident that participants frequently made comments that could be regarded as redundant or repetitive. However, the researcher considered that these comments may have been important to the study as they indicated a focus on particular aspects. Therefore, such comments were left in the transcription. Each transcription was then printed and proofread by the researcher by comparing the written document with the relevant audiotape. Transcription errors due to misinterpretation or misunderstanding of terminology were amended where possible. Where the comments on the tape were not clear, the notation 'inaudible' was made on the transcript and the information was not used for analysis.

The researcher made minor amendments to the transcriptions for a number of reasons. Where airline companies, personnel, or specific computer software brand names were identified, notations such as [airline name], [airline person], or [computer software] were inserted. To avoid confusion between a flight number used in the simulation (e.g., 705) and a time (e.g., 7.05am), all transcribed times were standardised to the 24-hour clock (e.g., 0705). Where abbreviations for airline ports or other specific terminology were inconsistent with the international standards, corrections were made. For example, for the airport of Canberra, Australia, the transcribed word 'CAN' (the actual code for Guangzhou, China) was changed to the correct code 'CBR'. Where terminology differed from airline to airline, the terminology used in the simulations was adapted for the participants in that airline. For example, in one airline the term 'tranship' (passengers who connect from one aircraft to another aircraft) was not as familiar to participants as the term 'connector'. Hence, the researcher referred to these passengers as connectors for the purpose of the data collection in that airline.

Reliability of the Data

One problem with data collected through qualitative methods is that it is invariably unstructured and unwieldy, as a result of verbatim transcriptions of interviews. Capturing the complexity of reality and making sense of the data then, is really dependent on their future use. The verbal protocols were analysed using protocol analysis procedures developed by Ericsson and Simon (1993). To enable this, each participant was allocated a unique case number which ensured the data gathered during the simulations matched the respective participant. The identification of main themes from the data required some interpretive license. To ensure that these interpretations did not misrepresent the data, judgments were sought from an expert who assisted in the design of the simulations. Participants' comments in relation to situation awareness and decision considerations were classified by the researcher according to three categories: *elementary*, *core*, and *advanced*. The comments were further classified into sub-categories to provide greater depth of information and focus the analysis. To provide measures of concurrence and to ensure consistency and rigour in the classification of the data, inter-rater comparisons were performed by the expert using a card sorting procedure. The expert sorted a series of cards containing the sub-categories of comments according to the three categories. Inter-relater comparisons were also conducted to classify participants' comments within each of the three categories. For these comparisons, the expert coded a random selection of transcripts (n=3) from the international simulations (N=33) and a random selection of transcripts (n=2) from the domestic simulations (N=19).

The inter-rater reliability estimates were calculated according to a formula provided by Goodwin and Goodwin (1985: 7): 'number of coding agreements/ number of coding agreements plus the number of coding disagreements', where agreement is the number of times the raters concur on the classification of a response. According to Miles and Huberman (1994), a 70 per cent inter-rater reliability is considered satisfactory. The overall mean inter-rater reliability for classifying sub-categories into the three categories was .73 and the overall mean inter-rater reliability for classifying participants' comments into the sub-categories was .88. Where there was disagreement between the researcher and the expert with regard to the classification of sub-categories or observed responses within the categories, the researcher and expert discussed each occasion and reached a consensus.

Methods for Recording Responses

To record participants' responses, various methods and devices may be used including written, tape-recorded, computer data entry or a combination of these approaches. However, the method used may influence the results. For example, having participants respond in writing may be tedious and limiting. One such study concluded that the artificialities of the paper and pencil used in a task may have influenced the results (Adelman, Tolcott and Bresnick 1993).

Researchers may also experience difficulty recording participants' responses due to the concentration involved in observation. A handwritten record in writing may be difficult to create and then analyse as notes may only have been selective verbalisations or may have paraphrased or omitted information. As a means of overcoming problems such as these, an audiotape recorder is sometimes used to offer a more reliable means for gaining a literal record of conversation. The recording can be replayed for clarity or to improve the transcription that is made from the taped comments. Importantly, a tape recorder also requires less attention, enabling the observer to focus on the task. Berg (1995) suggested that placing a tape recorder between participants may be disruptive. However, participants may become so involved in the task that they soon become accustomed to the presence of a taping device. In the study, participants were asked to verbalise their thought processes during decision-making. Audio-taping their comments provided an appropriate way of capturing the original data in its entirety. The experience of the researcher and his familiarity with the OCC environment, controller work practices and terminology commonly used were important factors for interpreting the observations. The use of think-aloud protocol in conjunction with simulations is a technique used commonly to gain insight into thought processes. Hence, this approach was also appropriate for the study given the difficulty of examining ways in which participants form mental pictures of situations.

Participants in the Study

Personal demographic data were collected for a range of variables including participants' gender and age and professional demographic data were collected including the participants' length of service in the airline industry, length of service in the Operations Control domain, and previous occupational experience.

Personal Demographic Variables

Gender Table 6.1 presents the frequency and percentage frequency distributions of participants classified by gender for international and domestic simulations. Nearly all participants (90 per cent) in the study were male. Anecdotal evidence gained from the expert panel suggests that the proportion of males to females of OCCs in the study is representative of airline OCCs.

Table 6.1 Frequency and percentage frequency distributions of participants classified by gender for international and domestic simulations. (*N*=52)

Gender	Participants in international simulations (*n*=33)		Participants in domestic simulations (*n*=19)		Total participants	
	f	%	*f*	%	*f*	%
Male	29	88	18	95	47	90
Female	4	12	1	5	5	10
Total	33	100	19	100	52	100

Age Table 6.2 presents the frequency and percentage frequency distributions of participants classified by age for international and domestic simulations. The low percentage of participants aged from 21–30 years suggests that OCCs may seek to employ controllers who are older and have considerable work experience. Nearly half the participants (42 per cent) were aged from 41–50 years of age. This may indicate an optimum age group for controllers.

Table 6.2 Frequency and percentage frequency distributions of participants classified by age for international and domestic simulations. (*N*=52)

Age	Participants in international simulations (*n*=33)		Participants in domestic simulations (*n*=19)		Total participants	
	f	%	*f*	%	*f*	%
21–30 years	1	3	4	21	5	10
31–40 years	13	39	4	21	17	33
41–50 years	13	39	9	47	22	42
51–60 years	6	19	2	11	8	15
Total	33	100	19	100	52	100

Professional Demographic Variables

Length of service in the airline industry Table 6.3 presents the frequency and percentage frequency distributions of participants classified by years employed in the airline industry for international and domestic simulations. Over half (58 per cent) of the participants had been employed in the airline industry for more than 15 years. This level of industry experience was evident for participants in both international and domestic OCCs and suggests that industry experience may be a key criterion for employment in OCCs.

Table 6.3 Frequency and percentage frequency distributions of participants classified by years of experience in industry for international and domestic simulations. (*N*=52)

Years in industry	Participants in international simulations (*n*=33)		Participants in domestic simulations (*n*=19)		Total participants	
	f	%	*f*	%	*f*	%
< 1 year	0	0	3	15	3	6
1–5 years	2	6	1	5	3	6
6–10 years	6	18	2	11	8	15
11–15 years	6	18	2	11	8	15
> 15 years	19	58	11	58	30	58
Total	33	100	19	100	52	100

Length of service in OCC Table 6.4 presents the frequency and percentage frequency distributions of participants classified by years in OCC for participants working international and domestic OCCs. Around one third of participants (31 per cent) in both international and domestic OCCs had been employed in OCCs from 6–10 years.

Table 6.4 Frequency and percentage frequency distributions of participants classified by years of experience in OCC for international and domestic simulations. (*N*=52)

Years in OCC	Participants in international simulations (*n*=33)		Participants in domestic simulations (*n*=19)		Total participants	
	f	%	*f*	%	*f*	%
< 1 year	5	15	4	21	9	17
1–5 years	8	24	3	16	11	21
6–10 years	10	31	6	31	16	31
11–15 years	2	6	3	16	5	10
> 15 years	8	24	3	16	11	21
Total	33	100	19	100	52	100

Previous occupation Table 6.5 presents the frequency and percentage frequency distributions of participants classified by previous occupation for international and domestic simulations. Half of all participants (50 per cent) in the study had acquired experience in airline crew scheduling and despatch departments prior to working in OCCs. This suggests that OCCs recognise the importance of employing controllers who have experience handling day to day operational situations in roles closely related to OCCs. Of the 33 participants who took part in the international simulations, 28 participants (85 per cent) indicated previous experience in aviation-related roles such as despatch, crew scheduling, pilot or flight engineer, reservations/sales, defence forces, customs, and schedules planning. The other five participants (15 per cent) indicated previous non-aviation occupations within finance, tele-sales and banking industries. Of the 19 participants who took part in the domestic simulations, 14 participants (74 per cent) indicated previous experience in aviation-related roles such as despatch, crew scheduling, pilot or flight engineer, reservations/sales, air traffic control (ATC), and defence forces. The other five participants (26 per cent) indicated previous non-aviation occupations within hospitality, police, automotive, and cinema industries. These data suggest that OCC management recognise the relevance of prior aviation experience as important for employment as a controller. Further, it appears that prior operational experience of day to day problem solving in areas closely related to the OCC is highly regarded by OCC management.

Table 6.5 Frequency and percentage frequency distributions of participants classified by previous occupation for international and domestic simulations. (*N*=52)

Previous occupational backgrounds	Participants in international simulations (*n*=33)		Participants in domestic simulations (*n*=19)		Total participants	
	f	%[a]	*f*	%	*f*	%
Despatch/airports/ Cargo	10	30	4	21	14	27
Crew Scheduling	6	18	6	32	12	23
Pilot/Flight Engineer	6	18	2	11	8	15
Reservations/sales	3	9	4	21	7	14
Defence	4	12	2	11	6	12
ATC	0	0	4	21	4	8
Schedules Planning	2	6	0	0	2	4
Customs	1	3	0	0	1	2
Non-aviation background	5	15	5	26	10	19

[a] Participants reported several previous occupational backgrounds. Therefore, percentages are not cumulative.

In summary, nearly all participants in the study were male, and most were between 31 and 50 years of age. Nearly three quarters of participants had more than ten years' aviation industry experience and nearly one-third had more than ten years' OCC experience. It appears that OCCs employ controllers with previous experience in specific operational areas closely related to OCCs such as despatch, airport handling, crew scheduling, pilots and air traffic control as nearly three quarters of participants had previous experience in these areas.

Summary

This chapter examined the research design and methods used in the study. The chapter explained that an inability to gather adequate data using survey questionnaires resulted in the abandonment of the preliminary study. The subsequent re-design of the study led to the use of simulations to examine decision-making processes of controllers in OCCs during disruptions. This approach gave the study a more appropriate means for examining decision-making processes of controllers during scenarios that replicated real-life OCC disruptions. Think-aloud protocol was used in conjunction with the simulations to capture thought processes of participants during decision-making and the method for recording and interpreting the captured data was explained. The expansion of the study to a multiple case method enabled comparisons to be made between decision-making processes in international and domestic environments. Finally, the chapter provided personal and professional demographic information about the participants involved in the study.

Chapter 7
The International Simulations

Introduction

A key focus of this study was to examine the influence of time constraints on the decision-making processes of controllers in OCCs. This was achieved by examining the decision-making processes in international OCCs and domestic OCCs separately. The reason for this is because decision-making in each environment is quite unique. In international operations, long-haul flights may be up to 18 hours in duration and even short-haul flights may be several hours long. The time an aircraft spends on the ground between flights (turnaround or turn time) is usually several hours as well. This enables a degree of freedom in terms of the time needed for controllers to evaluate disruptions that may arise and respond accordingly. However, in domestic operations substantial differences exist. Flights are often short (1–2 hours) in duration, aircraft are usually turned within 30–60 minutes and subject to the nature of the airline's operational network, there may be a myriad of crew and passenger connection times between inbound and outbound flights. Thus, decision-making in disruptions may be intense and decision time-frames are markedly shorter. Chapter Seven presents the findings and analysis in relation to decision-making processes by the 33 participants in the international simulations. Chapter Eight will present the findings and analysis in relation to decision-making processes by the 19 participants in the domestic simulations.

First, the scope of the analysis and the procedure used to examine the data in the international simulations are discussed. Second, the extent to which participants gained initial situation awareness in the simulations prior to the commencement of the three scenarios is discussed and information completeness is examined as an aspect of situation awareness in this part. Next, both the influence of experience and expertise on situation awareness and the way in which participants identify decision considerations as part of the decision-making process are examined. A major part of this chapter is the presentation of findings and analyses from the three international scenarios. Due to the complexity of the data overall, the findings and analysis for each scenario have been presented separately. The relationships among situation awareness, information completeness, and decision considerations are examined. In addition, the study investigates the influence of both participants' expertise and decision-making styles on decision considerations. Finally, the extent to which participants generate decision alternatives is examined.

Scope of the Study

The focus of the study was not to examine or assess the actions that participants might take to resolve a situation. This is because problems in OCCs typically have several solutions, all of which might make operational sense. Thus, examining the decision-making process was limited to the stage at which alternative courses of action in response to a problem may be generated. In terms of the rational models, this stage occurs immediately prior to the choice stage. In terms of the RPD model the stage is prior the point at which mental simulation of action occurs.

Conducting the International Simulations

The international simulations were conducted with 33 participants and consisted of two stages. The first stage required participants to gain awareness of the international flight schedules. The analysis of data for this stage examined the relationships among the variables' situation awareness and information completeness. The variables were analysed in relation to participants' experience and expertise. The second stage required participants to assess the situations presented in three individual scenarios and explain their decision-making processes. The analysis of data for the second stage examined the relationships among the variables: situation awareness, information completeness, decision considerations, decision-making styles, and generation of decision alternatives. These variables were analysed in relation to participants' expertise. During the simulations, all participants' comments were audio-taped, then transcribed verbatim. Appropriate passages from these transcriptions were used throughout the analysis to illustrate particular points, with the participant's case number placed in brackets following each passage.

Initial Situation Awareness

In airline OCCs, controllers gain initial situation awareness at the commencement of their shift by familiarising themselves with the flight displays and by receiving a briefing from other staff. Although it is uncommon for controllers to begin a shift with no actual or potential operational disruptions taking place, the simulations were designed such that no disruptions occurred during this familiarisation stage. This was important to the study as it enabled the researcher to maximise the collection of initial thoughts of participants.

Participants were given several minutes to view the flight schedules, describe their observations, and ask for any information they considered necessary. They were requested to verbalise their thoughts. In this way, their initial situation awareness could be identified. To prompt comments during this stage, participants were asked questions such as 'what do you start looking for?', 'what stands out?', 'how do you brief yourself?' and 'are you looking for particular things?' Asking these types of questions during the familiarisation stage was important for a number of reasons. This stage provided the first opportunity to capture participants' initial thoughts. In addition, the researcher could clarify that participants understood elements of the flight display and could explain any misconceptions they had. For example, it was important that participants understood the colour legend used in order to help them process relevant information.

Situation awareness was measured by recording comments made by participants during the familiarisation stage. While verbal protocols may not have given a complete picture of an individual's situation awareness, they provided rich information that was captured as it occurred. A few participants (12 per cent) commented that they used the familiarisation time to gain a general overview of the schedule. For example:

> [I look for] just a general overview of the state of the network and the passenger load [to see] if everything's full or [whether] we have some light loads today on certain sectors – that sort of overview. (217)

> So basically it gives you an overall view of how the day is going to unfold, so when something does unfold, then you sit back and wait. The advantage is that when the problem does happen, you're already prepared. (503)

Comments of other participants (nine per cent) indicated that they appeared to gain this overview by forming a picture of the situation during this familiarisation stage. This process has been identified in other studies. For example, in emergency management, senior medical despatch staff have referred to situation awareness as gaining a 'picture in the head' (see Blandford and Wong 2004: 435) and in ATC, controllers have formulated pictures as mental representations of the airspace (see Garland, Stein and Muller 1999). These studies suggest that forming a picture of the situation enables people to gain situation awareness and with it, the ability to facilitate decision-making. So it seems that forming a suitable picture of the flight display in an OCC would assist controllers in building their situation awareness, which could be important for facilitating decision-making in the three scenarios.

Most participants (85 per cent) indicated that they scanned the aircraft utilisation in quite considerable detail searching for potential problem areas. For example:

> [I look for] gaps, spare aeroplanes, maintenance restrictions – possibly ... any late running [operation] that's currently there, [and] where you can eliminate that by switching [aircraft]. (306)

> First of all I would look at whether there is any conflict of schedule. If there is any overlapping (a term used by controllers to signify when there is insufficient ground time between flights such that a delay will occur) then clearly there is a problem. (505)

Further, some participants (24 per cent) identified critical operations such as very long-haul flights which might imply operational restrictions. For example:

> Well, the long-haul flights would be affected by the flying time, the take-off weight [and] the crew. If the passenger loading is advanced and the flying time is too long and the wind is a headwind, [there will be] some restrictions on the flight. (504)

These comments demonstrate that participants appear to use a range of approaches to gain situation awareness. While it was evident that participants observed many aspects of the flight display, a further consideration of the study was to establish the range of information they sought.

Information Completeness

During the familiarisation stage, it became evident that many participants considered the information they observed as insufficient for gaining comprehensive situation awareness. These participants enquired further about information that would help them increase their knowledge of the airline operations to gain a more complete picture. In the study, this has been referred to as information completeness. In addition to gauging the information participants observed, it was also important to establish what information participants could not observe, but wished to know. For example:

> ... you'd [also] be thinking about the same thing in relation to weather, or any known industrial or environmental factors that could affect the performance of the schedule which could throw the hours limitation [out]. (308)

Although the flight display showed certain information such as maintenance requirements and limitations and passenger loadings on each flight stage, participants indicated that they needed additional information to develop situation awareness. For example:

> ... if I have anything which I think is in any way contentious, the first path would be Maintenance. I would have a quick conversation with the Maintenance planners or schedulers [for] whatever the particular relevant aircraft fleet is. (211)

> ... I would be interested in [information about] the airport as well ... which airport the aircraft is going [to], or what time the aircraft [is due to] arrive at [this] airport because of the curfew, runway works and delays. (504)

Most enquiries related to maintenance (42 per cent of participants), passenger loadings and tranships (36 per cent), regulatory (30 per cent), crewing (21 per cent), and weather (21 per cent). These concentrations suggest that many participants regarded additional, specific information such as the existence of curfews and slot times as vital for gaining situation awareness. In their normal workplace, either these participants are accustomed to being provided with more information, or they may go to some effort to obtain it themselves. In any case, these participants appeared dissatisfied with the information gained from their initial observation of the flight display and sought certain, additional information to address specific aspects of the operation.

Classification of Situation Awareness Comments

Endsley (1995a: 36) explained that the first level in achieving situation awareness is to ... 'perceive the status, attributes, and dynamics of relevant elements in the environment' and provided illustrations of this level of identification such as a pilot's perception of an aircraft, or of a mountain. However, within the first level of the model, she did not distinguish *between* elements that could be identified. For example, a pilot's perception of another aircraft is not given a different level of importance from the pilot's perception of a mountain. Rather, both of these perceptions appeared to be treated equally at the first level in her model.

In this study, the comments made during the familiarisation stage indicated that the participants considered a wide range of information in order to gain situation awareness. For example, some participants observed basic information such as flight loadings, while others identified more complex information such as critical, long-haul operations. This suggests that within the perception level of situation awareness described in the model, there may be several sub-levels. While controllers need to perceive relevant elements, the findings here are important to the study as they indicate that participants may consider some elements to be more relevant than others. If this is the case, it may be expected that perception of more complex elements would lead to a higher state of situation awareness. Thus, decision-making should become more effective. To explore this further, these sub-levels of situation awareness and the relationships among these sub-levels and participants' levels of experience and expertise were examined.

The participants' comments made during the familiarisation stage of the simulations were weighted and classified according to the degree of complexity into three categories of situation awareness. These categories were termed elementary, core, and advanced situation awareness. Within each category, sub-categories related to the focus of the comments. The inter-rater reliability for classifying participants' comments into the categories of situation awareness was .77. The inter-rater reliability for identifying comments within the sub-categories was .69. There were a total of 157 comments for situation awareness which were classified as elementary (20 per cent), core (55 per cent), and advanced (25 per cent).

Elementary Situation Awareness

Comments were classified as elementary when participants' observations appeared to be based on information readily obtained from viewing the display. Some general comments related to the way in which the participants used the physical display. For example:

> [The time line] adjusts me to the time of day I start and it just keeps in my mind the flights that are coming up ... [in case] I might get a phone call. I might get, you know, pilots checking in or whatever. (305)

The passenger loadings (booked passengers shown on each flight) were written clearly on the flight blocks. Over half the participants (55 per cent) commented that they scrutinised these loadings to ascertain which flights were heavily or lightly loaded with passengers. For example:

> [I am looking for] how many people are booked on the flight [and] whether it's exceeding the capacity of the aircraft. (213)

Comments were also classified as elementary when participants regarded the observed information as sufficient for the familiarisation stage. They did not seek further information either from the flight display or from the researcher during this stage. Further, they indicated that this level of awareness was sufficient until an operational disruption subsequently changed the situation. For example:

> Well, if you take over and there are no problems ... I wouldn't be looking for anything else until a particular problem occurred. (205)

These comments indicate that some participants requested information that would provide them only with a fundamental level of awareness. However, this degree of situation awareness may not, in itself, be sufficient for effective decision-making in complex situations. Comments at the elementary level of situation awareness were classified further into three sub-categories, according to the focus of the comments. Table 7.1 presents frequency and percentage frequency of comments classified within the elementary level of situation awareness.

None of the participants in the study identified sub-categories solely at the elementary situation awareness level. While most participants (75 per cent) identified elementary level sub-categories, 44 per cent of these participants also identified core level sub-categories and 56 per cent of them identified sub-categories at all levels of situation awareness. This suggests that the majority of participants realised that they could gain only a *superficial level* of situation awareness from information at the elementary level, and needed more advanced information to achieve a more comprehensive picture of the operations.

Table 7.1 Classification of comments within the elementary level of situation awareness. (*N*=33)

Sub-categories identified by participants	f	%
General overview only	9	29
Passenger loadings	18	58
Wait for disruptions to occur	4	13
Total number of comments	31	100%

Core Situation Awareness

The second category of situation awareness was termed 'core'. Comments classified as core were beyond the superficial comments evident at the elementary level. At the core level, participants identified sub-categories that could provide a more enhanced level of awareness that may enable the participants to anticipate potential disruptions. For example:

> [In] the utilisation I am going to be looking at, [I look for] ground time on the short-haul fleets. [The short-haul fleet is] generally a more dynamic environment so I will be looking at what type of ground time I have available to me; in other words, what type of hours I have available to me. (211)

> Well, you'd be looking at your schedule; whether it is on time and whether any of these flights are going to have overlaps or possible aircraft changes. (309)

Participants also identified more complex aspects of the flight display. For example, 79 per cent of participants referred to maintenance aspects, irrespective of the maintenance information already provided on the flight display.

> I look for where these maintenance requirements are ... so I would still have to assume that that [aircraft] has actual maintenance going to be done in those times, [and] small checks when [it] comes [back to base]. Most aircraft require some sort of checks. I'd have to know how much ground time those [checks need]. (216)

> I look at the aircraft that have very minimum hours to go, so if they have a check, you have to bring them back to base. That means you cannot leave it in the outstations ... you have to bring it back to base for a check. (404)

Comments at the core level of situation awareness were classified further into five sub-categories, according to the focus of the comments. Table 7.2 presents the frequency and percentage frequency of comments classified within the core level of situation awareness.

Table 7.2 Classification of comments within the core level of situation awareness. (*N*=33)

Sub-categories identified by participants	f	%
Aircraft schedules and patterns	28	32
Gaps between flights	20	23
Maintenance requirements or problems	26	30
Crew connections and duty limitations	12	14
Minimal passenger tranship times	1	1
Total number of comments	87	100%

The data in Table 7.2 indicate that the most relevant aspects of the flight display to participants were aircraft schedules and patterns (identified by 85 per cent of participants), maintenance (79 per cent), and gaps between flights (61 per cent). The focus by participants on these sub-categories suggests that they searched for information that might help them identify areas of potential disruption and likely consequences. This is in contrast to the focus on achieving an overview or observing basic information only, which was evident at the elementary level. Whereas 75 per cent of participants identified elementary level sub-categories, all participants identified one or more sub-categories at the core level.

It appears from this result that all participants recognised that fundamental information alone was insufficient to prepare them for operational disruptions. However, this does raise another point concerning the participants who had indicated core level sub-categories but had not mentioned any elementary sub-categories. A possible explanation for this is an oversight by these participants to verbalise all observations, despite probing questions to elicit further information or clarify comments.

Advanced Situation Awareness

The third category of situation awareness in the study was termed ‘advanced’. Beyond any awareness gained at the elementary or core levels, comments classified as advanced indicated that participants identified more critical and perhaps complex aspects which could affect certain flights. Many participants identified regulatory restrictions such as curfew hours (36 per cent) and current weather conditions or potential weather problems (30 per cent). Other participants (24 per cent) identified critical operations such as very long-haul flights. For example:

> Another thing ... is [an] aircraft performance depreciation situation. Sometimes for a few flights like long-haul London to Pacific ... if the flight hasn't departed,

> the aircraft will choose different heights ... the fuel saving can be ten to fifteen per cent ... so we have to choose the aircraft well. (403)

Several participants (24 per cent) indicated that not only did they look for flights which may become disrupted, but they were also searching for likely solutions. For example:

> ... so I'd look at one aircraft type to identify their routes and subsequently the passenger load [as well as] flights that I could possibly disrupt if I need to. [Then I would] do the same for the short-haul [flights]. When you start a shift [you] identify the likely problems and the opportunities to solve those problems – so we'd use this sort of sense to identify the areas that will cause you problems and related problems with operational constraints. (503)

These comments indicate that some participants looked well beyond the information contained within the display. By identifying limitations and seeking information with regard to airport restrictions, aircraft performance capabilities, and other operational constraints, participants revealed the importance of information for the decision-making process. Comments at the advanced level of situation awareness were classified further into five sub-categories, according to the focus of the comments: critical operations; potential for change/flexibility; potential weather problems; market conditions; and regulatory constraints. Table 7.3 presents the frequency and percentage frequency of comments classified within the advanced level of situation awareness.

Although all participants identified core level sub-categories, more than half of these participants (55 per cent) also identified sub-categories at the advanced level of situation awareness. This suggests that while some participants regarded core level sub-categories as sufficient for gaining satisfactory situation awareness, some participants sought an even greater level of awareness.

Table 7.3 Classification of comments within the advanced level of situation awareness. (*N*=33)

Sub-categories identified by participants	f	%
Critical operations	8	20
Potential for change/flexibility	8	20
Potential weather problems	10	26
Market conditions	1	3
Regulatory constraints	12	31
Total number of comments	39	100%

The conclusion drawn from the above analysis alludes to the considerable variation in the level of situation awareness gained by participants. It is evident that some participants seek to gain only a basic level of awareness. In contrast, other participants seek an enhanced level of situation awareness by identifying information likely to help them anticipate disruptions and provide solutions.

Experience

To investigate whether the variations in levels of situation awareness were influenced by participants' experience, the study investigated the relationship between situation awareness and experience. In the international simulations, 67 per cent of participants had gained experience in crew scheduling, maintenance, fleet scheduling, or airport functions. The comments of several of these participants (15 per cent) indicated that they familiarised themselves with aspects of the simulation specifically related to their own backgrounds. For example:

> ... [I look for] what delays have been incurred and basically from a crewing point of view to see if a flight has been delayed – whether the crew actually have arrived and reported [for duty]. [I also look to see if] all the flights [have] got their required number of crew on board the flight. Once I am in the chair, [I] look at what different scenarios I can have with the crew. (302)

> Well, I'm a maintenance person anyway and I'm looking for opportunities to do maintenance. I'm looking for opportunities whereby we can catch up [with maintenance work]. For each of the aircraft – one of the things that needs to be done is to tailor what aircraft do what flights according to the hours available and the work you want done at the end of the day. I also notice that [maintenance] work is being done at Pacific. (307)

> [The other thing] is knowing your market, [so] that if we're full, the other carriers are full; if we're empty, they're likely to be empty. [This comes from] experience for having worked in airports and working in them from an airport point of view. (308)

These comments suggest that specific past industry experience contributes to the way in which some participants accumulate situation awareness. This finding is consistent with previous research which suggests that the acquisition of situation awareness is influenced by an individual's abilities, experience, training, and increasing responsibilities (e.g., Blandford and Wong 2004, Kass, Cole and Stanny 2007), but it also emphasises the importance of specific industry experience. In particular, the finding raises the question as to whether these participants fail to familiarise themselves with other equally relevant information. To investigate the relationship between situation awareness and experience, the study examined participant experience at two levels: years of industry experience years of OCC domain experience. Experience was examined according to a scale which classified levels of experience into five categories: less than one year's experience, one to five years' experience, six to ten years' experience, 11 to 15 years' experience, and greater than 15 years' experience. Classifying experience levels according to this scale enabled the relationship between experience and situation awareness to be examined to a high degree facilitating in-depth comparisons and analyses of the data. The frequencies of participants' comments were analysed according to years of experience by comparing the mean numbers of comments, where the mean (*M*) was determined as the number of comments in a category divided by the number of participants in the category who made the comments. The mean numbers of comments per participant for situation awareness were classified according to experience in industry (Table 7.4) and experience in OCC domain (Table 7.5).

Several patterns emerge from the data in Table 7.4. First, there is a high concentration of responses at the core level regardless of years of industry experience. This suggests that participants may consider key, relevant information as vital for achieving a satisfactory level of situation awareness. Second, for core and advanced level comments of situation awareness, the mean numbers of comments are higher for participants with greater levels of industry experience which suggests that greater industry experience is likely to lead to enhanced situation awareness. In summary, it appears that industry experience may help to explain the variation in the amount of situation awareness gained by participants.

The relationship between situation awareness and OCC domain experience was also considered. Table 7.5 examines this relationship according to the number of years participants had been in the OCC.

Table 7.4 Frequencies and mean numbers of comments for situation awareness according to years of experience in the airline industry. (*N*=33)

	Years of experience in the airline industry							
Situation awareness	**1–5[a]** *n*=2		**6–10** *n*=6		**11–15** *n*=6		**>15** *n*=19	
	f	*M(SE)*[b]	*f*	*M(SE)*	*f*	*M(SE)*	*f*	*M(SE)*
Elementary	3	1.5(.25)	5	.83(.14)	6	1.0(.22)	17	.89(.19)
Core	5	2.5(.25)	14	2.3(.18)	15	2.5(.30)	53	2.8(.25)
Advanced	-	-	7	1.2(.35)	10	1.7(.58)	22	1.2(.32)

[a] There were no participants with less than one year of industry experience.

[b] *M* = the mean number of comments per participant in the category.

SE = estimated standard error of the mean.

Table 7.5 Frequencies and mean numbers of situation awareness comments according to years of experience in OCC. (*N*=33)

	Years of experience in OCC									
Situation	**<1 year**		**1–5**		**6–10**		**11–15**		**>15**	
awareness	***n*=5**		***n*=8**		***n*=10**		***n*=2**		***n*=8**	
	f	*M(SE)*[a]	*f*	*M(SE)*	*f*	*M(SE)*	*f*	*M(SE)*	*f*	*M(SE)*
Elementary	7	1.4(.51)	6	.75(.25)	10	1.0(.15)	1	.50(.50)	7	.88(.23)
Core	12	2.4(.24)	21	2.6(.32)	27	2.7(.26)	5	2.5(.50)	22	2.8(.49)
Advanced	4	.80(.49)	6	.75(.25)	18	1.8(.49)	1	.50(.50)	10	1.3(.73)

[a] *M* = the mean number of comments per participant in the category.
SE = estimated standard error of the mean.

The data in Table 7.5 indicate a high concentration of comments at the core level regardless of years of OCC domain experience. A similar pattern was observed for the core level comments for experience in industry. This suggests that participants consider that key relevant information is important for achieving a satisfactory level of situation awareness regardless of the number of years of industry or OCC domain experience. In some categories, very low frequencies of comments were evident. Therefore, caution has been exercised in interpreting these results in terms of the study and with regard to generalisability of the results. In relation to core level situation awareness, the means for participants with greater than 15 years' OCC domain experience are higher than the means for other participants. This suggests that considerable OCC domain experience may help participants gain situation awareness by identifying many aspects of the flight display and in particular, more complex aspects such as critical operations. In summary, it appears that participants' OCC domain experience may help to explain the considerable variation in the level of situation awareness gained by participants.

The data in Tables 7.4 and 7.5 indicate that participants with less than five years' industry experience and less than one year OCC domain experience identify more elementary aspects of situation awareness than other participants. However, it appears that with increased years of experience, both in industry and in the OCC domain, participants identify more complex information. While the importance of domain experience for gaining situation awareness has been emphasised in other research, the results from this study suggest that both industry and OCC domain experience may be important for determining the way in which participants gain situation awareness.

Expertise

This section examines the relationship between situation awareness and expertise. In Chapter Five, it was established that experts have extensive knowledge of their domain and this knowledge is acquired from many years of experience. The domain in the study is the OCC. Despite numerous definitions of expertise, there is considerable support for the view that people may become experts in their domain only after ten years' experience. Consequently, ten years' experience was used in the study to differentiate between novices and experts. Table 7.6 presents the relationship between situation awareness and expertise for novices and experts according to their level of expertise in both industry and in OCC.

Several patterns emerge from the data in Table 7.6. First, in regard to the mean numbers of comments for each category of situation awareness: elementary, core, and advanced, there is a high concentration of responses at the core level regardless of the level of participants' industry or OCC expertise. This is consistent with the findings in Tables 7.4 and 7.5 and suggests that both novices and experts seek key relevant information to achieve a satisfactory level of situation awareness. Second, participants who were novices in both the industry and OCC had the lowest mean number of comments for participants in core and advanced categories of situation awareness. This may be expected due to their limited industry and OCC experience.

Table 7.6 Frequencies and mean numbers of situation awareness comments by novices and experts according to the level of expertise in industry and in OCC. (*N*=33)

Situation awareness	Novices in industry[a] and OCC		Experts in industry and novices in OCC		Experts in industry and OCC	
	n=8		*n*=15		*n*=10	
	f	*M(SE)*[b]	*f*	*M(SE)*	*f*	*M(SE)*
Elementary	8	1.0(.19)	15	1.0(.22)	8	.80(.20)
Core	19	2.4(.18)	41	2.7(.23)	27	2.7(.40)
Advanced	7	.88(.35)	21	1.4(.36)	11	1.5(.60)

[a] No participants were both novices in industry and experts in OCC.

[b] *M* = the mean number of comments per participant in the category.

SE = estimated standard error of the mean.

In contrast, the means for participants who were considered experts in industry were higher than the means of novices according to each category of situation awareness. Further differences were evident between participants who were considered novices in the OCC and those who were considered experts in the OCC. OCC novices appeared to focus on elementary sub-categories. However, OCC experts appeared to focus on advanced sub-categories. This finding is consistent with earlier findings in the study for participants with high levels of industry and OCC experience and suggests that the greater the level of expertise, the more likely participants are to build a more comprehensive level of situation awareness.

The conclusion drawn from these findings supports previous research that identified the importance of domain expertise for achieving situation awareness. However, the study found that industry expertise is also important. Therefore, if participants are expert in both the industry and the OCC, they are likely to achieve a more advanced level of situation awareness which should enable them to make more effective decisions.

Decision Considerations

An important part of the decision-making process is the generation and evaluation of alternatives. From these alternatives a choice is made and action can be taken. However, to arrive at the alternatives, controllers need to take several considerations of factors into account. While the decision-making models shown in Chapter Two depict several steps that lead to making a decision, none of the models appears to recognise the presence or importance of the considerations that may be taken into account prior to the generation of alternatives. Consequently, the study examined the considerations participants regarded as relevant to generating the alternatives. These were termed *decision considerations*. The patterns that emerged from participants' comments led to the classification of these comments according to the degree of complexity into elementary, core, and advanced categories.

Classification of Decision Considerations Comments

Comments were classified as elementary when participants considered fundamental aspects of the scenario. Further, participants based these considerations on information given to them without challenging the source or validity of information. For example, they may have begun to consider operational consequences of a situation before confirming that information provided came from a reliable source. Comments classified at the core level indicated that participants identified opportunities to reduce the consequences of a situation. For example, they might have questioned whether the delayed flight was likely to have implications for crew duty hours, or whether the flight may breach a curfew at an airport. Comments classified at the advanced level indicated that participants had considered ways to avoid the occurrence of a situation. For example, they might have identified an opportunity to substitute a replacement aircraft or crew, which negated the need to delay a flight. Decision considerations were examined for each of the three scenarios. As each scenario contained a unique operational situation, the decision considerations were specific to that scenario.

Analysis of Scenarios

Beyond gaining initial situation awareness of the schedules, the second stage of the international simulations required participants to consider the situations presented in three individual scenarios. The expert panel provided an optimum solution for each scenario and a set of decision considerations that participants should take into account during the decision-making process. This provided a means for evaluating participants' responses in the scenarios. The scenarios were analysed according to participants' comments in relation to their revised situation awareness, information completeness, decision considerations, levels of expertise, decision-making styles, and generation of decision alternatives.

Scenario One

The following initial briefing was provided to participants at the start of Scenario One:

> The time is 2100. Flight 703 operating from London (LHR) to Pacific (PCF) has been unserviceable in LHR. The aircraft has eventually departed two hours late and is picking up time into Pacific. The ETA (estimated time of arrival) in PCF is 2315. There is no crewing problem. All Tranships are OK, except a Ministerial Delegation of 55 connecting with Flight 741 to Melbourne (This flight is scheduled to depart at 2320). The Airport Manager at Pacific says he IS holding 741 and expects to get it away at 2345. (Subsequently, the inbound flight was delayed further necessitating further decision-making by participants).

Scenario One required participants to consider whether and how long they might hold Flights 741/742 from Pacific (PCF) – Melbourne (MEL) – Sydney (SYD) – Pacific to wait for tranship passengers from the delayed inbound flight from London (LHR). Delaying the outbound Flight 741 required participants to take into account several considerations such as the commercial impact (especially the Ministerial delegation of 55 passengers), crew hours, and a potential breach of curfew in Sydney as part of their disruption recovery strategies. Figure 7.1 below provides a graphical depiction of this scenario and the flight display can be viewed in Appendix A.

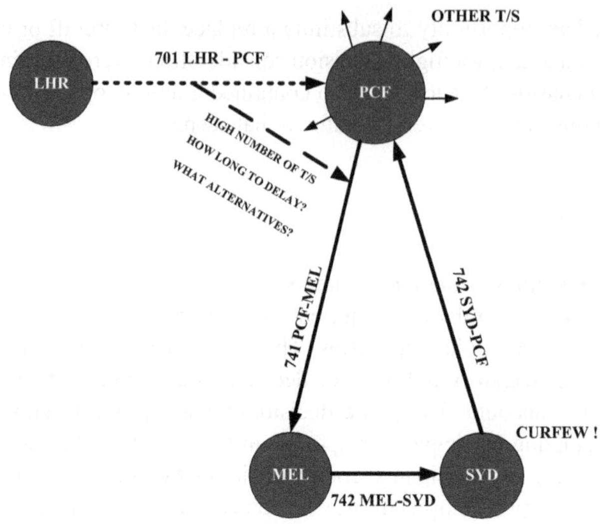

Figure 7.1 Graphical representation of international scenario one. See Appendix A for international utilisation (not to scale)

From several alternative actions that could have been considered, the optimum solution provided by the expert panel was initially to extend delays to the outbound flight from PCF until such time that the operation breached the curfew at Sydney at which point the operation could be re-routed to PCF-SYD-MEL-PCF. The inter-rater reliability for classifying participants' comments into the categories of decision considerations in Scenario One was .75. The inter-rater reliability for identifying comments within each category was .84.

Revised Situation Awareness

In the familiarisation stage of the simulations, participants gained situation awareness through observation of information that was evident in the flight display and by requesting further information. Participants enhanced their level of situation awareness in the course of the scenario briefings and by requesting specific information in relation to the scenario problems.

Information Completeness

Participants' comments related to both the reliability and source of information on which they could base decisions. For example:

> I'd check that the information is correct. Obviously from that point of view [I'd ask] what their revised estimated time of arrival is, if it is already in flight. And I would monitor that as it goes. (203)

> It's not just [sufficient] to work on block (scheduled flight) time because that wouldn't be absolutely accurate in today's operations. I think we need to look for actual times. (400)

These comments demonstrate the importance that participants placed on gathering sufficient and relevant information about a situation. This evidence-based approach to decision-making is a critical step in the decision-making process. In the study, it was difficult to determine how much information participants considered to be sufficient, as they requested further information throughout the scenario. However, people have limited information processing capabilities so they filter out some of it. For example:

> Sometimes [having] additional information is not the right timing. It just complicates everything. (401)

In contrast, some participants requested information several times in relation to one specific aspect such as a crewing situation for a particular flight. For example:

> How are we off for crew hours? ... can we extend the crew duty? ... Have we got any crews laying over? ... Have you got another crew? ... What about that inbound crew that [operated flight] 706? ... so the crew that have reported need 12 hours [rest] do they? (306)

In these circumstances, it was important to capture the variety of comments relating to that aspect. Therefore, multiple comments were recorded providing that the enquiries were not repetitive. In Scenario One, there were 192 requests for further information by participants. Most participants focused on information related to three main aspects: crewing (79 per cent of participants), airport handling (76 per cent), and passenger loadings and tranships (88 per cent). These aspects were far less prominent during the familiarisation stage which suggests that participants took into account scenario specific information in addition to the initial information. It was evident at the commencement of the simulation that participants regarded certain information as vital for acquiring an initial level of situation awareness. However, the comments made by participants during this scenario indicate that the acquisition of situation awareness appears to be a cumulative process.

Decision Considerations

Participants made 142 comments relating to decision considerations in Scenario One. The comments were classified according to elementary (20 per cent), core (63 per cent), and advanced (17 per cent) categories.

Elementary Decision Considerations

Comments were classified as elementary when participants appeared to acknowledge the information provided as sufficient for the scenario. The initial details provided in Scenario One informed participants that a request to delay a flight had been made by a commercial department in the airline. Many participants (52 per cent) readily accepted this information without challenging its source or credibility. For example:

> OK, first ... I am going to be looking at the impact that [flight] 741 being delayed will have on its arrival at Melbourne and its subsequent sector. (307)

> ... most of that delay is ... not the issue for me. The issue for me is the [late arrival] coming into Melbourne, where the aircraft is going to next and where it goes from there. (311)

Other participants (33 per cent) focused on potential transhipping (connection) problems as a result of the delayed flight. For example:

> Of course, with an hour delay, I would not think that the crew would be a problem. Of course we'd look into other passenger connections ... there may be passengers from [Pacific] to Melbourne. (405)

> Would you be able to tell me whether this delay of about an hour is going to give any connecting problems for the passengers going to Melbourne or Sydney? (502)

Most participants (73 per cent) identified elementary decision considerations. Further, participants appeared to accept the information provided in the scenario as a basis for investigating the consequences of the delayed flight. There was little evidence to suggest that participants challenged the information. A possible reason for this may relate to a design limitation of the simulations. Participants may have regarded the information given to them as absolute and may have been reluctant to question it.

Core Decision Considerations

Comments classified as core considerations indicated that participants had a greater comprehension of the consequences caused by delaying the flight. In contrast to comments at the elementary level, several participants (33 per cent) challenged some of the information they received as the scenario was introduced. For example:

> ... I have an input from the airport manager who is there and [is] the person on the spot for me to glean what is achievable and I would question his timings. (211)

> [A] 30 minute connection is what twigged as probably unachievable ... how accurate are we with an ETD (estimated time of departure)? The consequences of it, I'd be looking at next, but the question is, how accurate is [a connection time of] 30 minutes for 120 passengers? (211)

> ... even though the duty manager has told us we're not going to get away until 2345 ... he's only given us half an hour transit to get these 55 politicians [from one flight to the other] and knowing what a delegation's like, I think they'd need a longer time than 30 minutes. (217)

Most participants (85 per cent) identified problems that could disrupt the delayed flight further, such as curfew times, crew duty hours, crew flight time limitations, and airport handling issues. For example:

> Well now, that means you've got a curfew time. I mean you can arrive in Sydney obviously, but not depart because of the curfew. (403)

> So I would ask the airport [manager] again [as to] what's the best time they can get the passengers from [flight] 703 to [flight] 741, providing we can organise parking bay[s] next to each other to facilitate the transfer. (502)

Many participants (79 per cent) considered opportunities to mitigate the disrupted operation, such as shortening the scheduled ground times at airports, or asking if the flight could operate at a faster speed (high speed cruise) to recover some of the delayed schedule. For example:

> There isn't much of an opportunity to pick up any late running, so it's just a matter of ... asking all stations to do their best to get the aircraft turn-rounds as quickly as possible. (315)

> Generally we'd be looking at getting advanced speed crossings, if that was at all achievable on any of the sectors, to minimise the impact. (309)

These comments show that participants identified aspects well beyond those at the elementary level. They sought clarification of information, recognised potential constraints, and identified ways to minimise the disruption. This demonstrates a markedly higher level of comprehension of the situation. All participants identified decision considerations at the core level. However, 27 per cent of these participants did not indicate elementary level comments. This suggests that either they did not consider information such as tranships, or they did not verbalise the information, even if it had been considered. This demonstrates one of the difficulties of examining ways in which participants use information for decision-making, despite claims that verbal protocols enable this process to occur.

Advanced Decision Considerations

Comments classified as advanced considerations indicated that participants demonstrated thinking beyond the immediacy of the situation. For example:

> ... we might want to ask our representatives in Sydney, ahead of time, [to] try and find out from the relevant authorities [whether they will] consider giving us curfew dispensation to operate during curfew hours. (400)

> Instead of coming out for an 1130 UTC departure, it may be ... a 1300 departure, so we get the Crewing people to roll their crew time (delay the crew signing for duty) for the departures that we are going to set and what that does is save us any crewing problems. (201)

> So I start to look at things like, can we go [from] Pacific, straight to Sydney, then via Melbourne; in other words come through Sydney [first]. I appreciate the loads are full, so our ability to carry them through to Sydney and pick up the joining load may be quite hard, but that's the intention. (211)

> Well, the holding of the crew at home base means that until they've reported, their duty doesn't start. So it actually allows me more ability to utilise this crew and extend them further if, for any reason ... we have subsequently got another ... delay of some sort. It just gives me that additional time factor. (302)

These comments suggest that some participants appeared to anticipate the consequences of the situation and consider these in great depth. Whereas all participants made comments which were classified at the core level, only around half of the participants (58 per cent) identified considerations at the advanced level. Core level comments emphasised the way in which participants managed the disruption by working within operational limitations. For example, 61 per cent of participants focused on crew duty limitations that would be affected by delaying the flight.

However, at the advanced level, 33 per cent of participants reported ways to overcome the limitations, such as replacing a crew member or delaying their starting time for duty. This suggests that while most participants identified operational constraints, many of them did not necessarily seek ways to alleviate these constraints. Therefore, either these participants may not have identified remedies for the situations, or they may have been content to work within limitations they have identified. Hence, in OCCs they may only consider ways to *manage* rather than *prevent* an operational disruption. The results highlight deficiencies in current disruption handling and identify opportunities for training of controllers in OCCs.

The classification of participants' comments into the three categories of decision considerations demonstrated substantial variation in the way that participants considered the implications of the scenario. To investigate whether this variation was influenced by participants' levels of expertise, the study investigated the relationship between expertise and decision considerations.

Expertise

The relationship between expertise and situation awareness was examined earlier in this chapter. The findings showed that industry and OCC expertise were likely to influence the way in which participants gained situation awareness. This section examines the relationship between expertise and decision considerations. The analysis used ten years' experience to differentiate between novices and experts as previously established.

Chapter Five discussed differences in decision-making processes between novices and experts. In the study, differences in the ways that novices and experts identified decision considerations became evident from the participants' comments. Novices at both industry and OCC levels appeared to take few decision considerations into account. Their comments often were limited to asking for further information about an aspect they had identified. For example:

> ... I would check that ... and make sure that my tech[nical, pilot] crew hours are on [the trip plan and] that they're not going into disrupt[ion time] or anything. OK, so I go to my crewing person ... and what are [the crew's] hours [limitations]? OK, so that would be fine, crewing wise. (305)

Some novices seemed unsure as to what they might consider in relation to the situation. For example:

> Not a lot [going through my mind] at the moment. Where are we heading with this again? We're running ... (212)

Other novices appeared to identify several, separate aspects relating to a situation, often in rapid succession. These comments were interspersed with appropriate responses from the researcher:

> [In relation to] this aircraft ... [is the] maximum capacity ... 450? I'd just check whether there are any other alternatives that you can use. So the other option is ... we could upset the other 350 passengers. What I am concerned [with] is the different sectors ... We have a crew change in Melbourne too? OK, what is the FTI (Flight time interval)? How about the crew hours for this evening? They operate the two sectors. What is the flight time limitation on this? (500)

It was evident from the comments that these participants searched for various clues that might help them grasp the extent of the problem. These comments are consistent with other research which contends that novices only see bits and pieces. This may be because of the way that novices focus on the surface features of a situation. In contrast, experts appeared to adopt a far more comprehensive approach to the situation and considered how aspects of a situation influenced each other. For example:

> What you would need there [is to] probably hold the [flight] 742 crew by up to an hour if that was what the delay was ... with Sydney, you [will] possibly be cribbing curfew to get out, so with an hour delay, 2230, it is still on that, but if it went any later you have to start worrying about Sydney curfew. If it got really late and you weren't going to make Sydney curfew, well you might have to look at coming straight [from Pacific] to Sydney and [reversing the routing of the flight] ... going from Pacific to Sydney and then to Melbourne. (205)

In particular, the experts' comments demonstrated their comprehension of the complexity of the situation. As the scenario progressed, it became evident that experts anticipated consequences beyond the immediate situation. For example:

> He needs an hour on the ground, so that's now a two hour delay for that flight ... now that's getting a bit ugly ... making it a two hour delay here. So that pushes [the flight] out ... I'm just extending [the delay] on the chart. Now that really has put the cat amongst the pigeons there. Now [that] just does change how I thought about the previous [information]. There has to be a cut-off somewhere. I'm not willing to delay this thing indefinitely. (310)

Some participants indicated their use of a time optimisation strategy. For example:

> ... my point of view [with] that time factor is whether or not I can actually hold the crew at home base ... well, the holding of the crew at home base means that until they've [signed on for duty], their duty doesn't start, so it actually allows

> me more time; more ability to actually utilise this crew and extend them further ... It just gives me that additional time factor; another hour up my sleeve. (302)

This strategy of 'buying' time enabled the participants to assess the situation and defer the decision. The comments provide an insight into the ways in which novices and experts comprehend complex situations. They also indicate that experts appear to manage a disruption better than novices by identifying relationships between decision considerations, projecting likely consequences of a situation, and using strategies to ensure their decisions are likely to lead to appropriate actions. Table 7.7 presents the relationship between expertise and decision considerations for novices and experts according to their level of expertise in both industry and OCC.

Several patterns emerge from the data in Table 7.7. First, there is a high concentration of comments at the core level. This is similar to the findings describing the relationship between expertise and situation awareness and suggests that there may be a focal set of considerations on which all participants concentrate regardless of their levels of expertise. Second, participants who were experts at industry level but novices at OCC level identified the most core decision considerations. In addition, participants with both industry and OCC expertise identified the most advanced considerations. These findings suggest that industry expertise may enable participants to identify constraints and manage a situation within these constraints. However, the data also suggest that participants with expertise in both industry and OCC are more likely to anticipate situations occurring, enabling preventative action to be taken. The findings help to explain the relationships among decision considerations, industry expertise, and OCC expertise. In addition, the results highlight the importance of expertise in the identification of complex decision considerations.

The data presented in Table 7.7 also aid our understanding characteristics of experts, especially of ways in which they assess a situation quickly, see complex displays in terms of an overall picture, and use this representation to prepare actions in advance.

Table 7.7 Frequencies and mean numbers of decision considerations comments by novices and experts according to their level of expertise in industry and in OCC. (N=33)

Decision considerations	Novices in industry[a] and OCC		Experts in industry and novices in OCC		Experts in industry and OCC	
	n=8		n=15		n=10	
	f	$M(SE)$[b]	f	$M(SE)$	f	$M(SE)$
Elementary	7	.88(.23)	13	.87(.17)	8	.80(.20)
Core	16	2.0(.63)	47	3.1(.22)	26	2.6(.37)
Advanced	5	.63(.26)	9	.60(.16)	10	1.0(.26)

[a] No participants were both novices in industry and experts in OCC.

[b] M = the mean number of comments per participant in the category.

SE = estimated standard error of the mean.

Decision-making Styles

To investigate the influence of decision-making styles (DMS) in the OCC, the study examined the extent to which participants used rational and intuitive decision-making styles during the decision-making process in the international simulations. The analysis of data also examined the extent to which rational and intuitive decision-making styles appeared to differ between novices and experts.

Rational Decision-making Style

Rational decision-making has been described earlier as a logical, systematic process of analysis that occurs in a series of steps. In the study, the extent to which participants used a rational decision-making style (DMS) was examined by classifying their comments as low rational DMS or high rational DMS.

Comments classified as low rational DMS indicated that participants showed little evidence of using a step by step approach. For example:

> [I would] look for other airlines going down to Melbourne and put the passengers onto that ... if that's the case, then we'd check [that] the crews are OK and we'll need to know what sort of delay to get them across. (501)

> Based on those times there, eyeballing what we've got, without a calculator or anything like that ... it's going to blow out some time for the next day for the [flight] 741, so that's the same time as [another aircraft]. (213)

Comments were classified as high rational DMS when participants suggested a series of considerations in a highly systematic way. For example:

> Do we have another flight in Paris? [I'd] get the traffic staff to check other airlines to protect the passengers. [I'd ask] how many first and business [class passengers] ... and ask them to book [the passengers] in [to hotels] ... all the rooms near the airport. [Then I'd] send the crew [to the hotel] to [take crew] rest. (508)

> ... say that's going to be delayed an hour or so straight away. I would check the crew hours [and] see if they can cope with an hour's delay or not. I'd advise the [passenger handling people] who take care of the on-carriage (onward connections) ... [then] I'd go to my aircraft maintenance [controller] and check [the latest information about] this delay. (305)

> ... the first thing we'd check is that everything is OK to handle these two [flights] in both airports, Manila and Taipei. Then we'd have a chat with the crew [to see if] the crew hours [are] OK or not. If not OK, then we'd have to send crews from home base to pick up the aircraft to [go to] those two airports. Then we'd have passenger services put [the crews] on something special arranged. We would have to alert ground handling ... (504)

These comments demonstrate that for some participants, the use of a systematic process seemed minimal, while other participants appeared to use an elaborate series of steps. Table 7.8 presents the relationship between participants' use of rational DMS and expertise, according to the level of expertise in both industry and OCC.

Table 7.8 Classification of participants' use of rational decision-making style according to their level of expertise in industry and in OCC. (*N*=33)

Rational DMS	Novices in industry[a] and OCC *n*=8		Experts in industry and novices in OCC *n*=15		Experts in industry and OCC *n*=10	
	n	%	*n*	%	*n*	%
Low	2	25	4	27	3	30
High	6	75	11	73	7	70

[a] No participants were both novices in industry and experts in OCC.

The data in Table 7.8 show that most participants appeared to use high rational DMS regardless of their level of expertise. These findings are in contrast to previous research which suggests that rational decision-making in uncertain, dynamic environments is unlikely to be successful (e.g., Klein 2001). However, the time for decision-making in international operations is often extensive and Scenario One was the least complex of the three scenarios in the international simulations. Accordingly, the time for decision-making in this scenario was not restrictive.

Intuitive Decision-making Style

Intuitive decision-making has been described as being based on gut-feeling, or having a feeling as to what is right and what is wrong and is largely determined by a person's experience. In the study, it was evident that participants had experienced similar situations to those described in the scenario. For example:

> There'd be slot time implications out of Paris. Because of the European situation, you have to let flight despatch [and] ATC know [and] you have to obtain the nearest slot [take-off time] out of Paris for that. [For] anywhere in Europe, that is the case. (203)

> [Regarding the] Melbourne passengers, so [we have] half an hour to spare. [This is] a bit tight. Ah, normally we find out the schedule ... normally we give one hour to get the passengers across. With this running late, this time [we have] about half an hour for this transit – OK you'd get a delay to the flight. (403)

The extent to which participants used an intuitive decision-making style (DMS) was examined by classifying their comments as low intuitive DMS or high intuitive DMS. Comments classified as low intuitive DMS indicated that participants demonstrated limited use of gut feeling. For example:

> We'll leave the [flight] 741 as it is. We won't touch that yet ... [we'll] wait and see. There's no point jumping the gun. Let's just see what the final story is from the engineers I think. (214)

> It doesn't appear that there would be any maintenance problems with that running late. So I would ... just issue the re-schedule signal. (307)

Comments were classified as high intuitive DMS when participants appeared to use a more complex, intuitive approach to the situation. While they might have known what to do, they seemed to sense what would and would not work. For example:

> So, OK. It's fantastic he can do it, but I am still reticent to say that we are only copping a one hour delay. I would believe it would be more. I wouldn't want to start publishing one hour [which then might become a] rolling delay. I believe [that] the time ... published [should be] something achievable. What is an achievable time? (211)

> ... the duty manager has ... only given us half an hour transit [time] to get these 55 politicians [to transfer flights] and knowing what a delegation's like, I think they'd need longer than 30 minutes basically to get the [passengers across]. The ramp might be ok to get the bags ... off pretty quickly but I think to move a group of that size ... and get them on board [the connecting flight] ... might take a bit longer than that. (217)

The comments support previous research suggesting that intuition is characterised by having a sense of what will and will not work and being able to comprehend the situation holistically. All participants were given the same information with regard to the scenario. However, the comments above demonstrate considerable variation in the way participants used an intuitive decision-making style. Table 7.9 presents the relationship between participants' use of intuitive DMS and expertise, according to the level of expertise in both industry and OCC.

The data in Table 7.9 indicate that only about half of all participants appeared to use high intuitive DMS. There was little evidence to suggest that experts used intuitive DMS to any greater extent than novices in this scenario. This may be an expected result for Scenario One as the scenario was straight forward and considerable information was provided.

Table 7.9 Classification of participants' use of intuitive decision-making style according to their level of expertise in industry and in OCC. (N=33)

	Novices in industry[a] and OCC		**Experts in industry and novices in OCC**		**Experts in industry and OCC**	
Intuitive DMS	***n*=8**		***n*=15**		***n*=10**	
	n	%	*n*	%	*n*	%
Low	4	50	9	60	5	50
High	4	50	6	40	5	50

[a] No participants were both novices in industry and experts in OCC.

Generation of Decision Alternatives

Actual decisions made by OCC controllers may vary for a number of reasons. For example, there may be several ways to resolve a disruption and each of these ways may result in a satisfactory outcome for the OCC. The focus of the study was not to evaluate decision outcomes but to examine the decision-making process to the point of generating decision alternatives. Therefore, the decision outcomes were beyond the scope of the study.

In Scenario One, participants were required to determine whether to delay a flight to wait for passengers from another connecting flight. Although, such a delay may have resulted in exceeding a curfew limitation, nearly all participants (91 per cent) agreed to delay the flight. Several participants (21 per cent) who were all novices in industry and OCC did not indicate either an awareness of, or a concern for, the curfew. One appropriate and creative alternative was to re-route the flights (from PCF-MEL-SYD-PCF to PCF-SYD-MEL-PCF) thereby enabling the delayed flight to operate at an earlier time through Sydney due to its curfew. This alternative was a favoured solution to the problem and was designed into the scenario in consultation with the expert panel. Several participants (24 per cent) with industry or OCC expertise proposed this alternative as a viable means of overcoming several problems created by the delayed flight. This finding is consistent with earlier findings in this chapter suggesting that experts were more likely than novices to identify advanced decision considerations with a view to creating decision alternatives that overcome limitations.

Scenario Two

The following initial briefing was provided to participants at the start of Scenario Two:

> The time is 0655. Flight 705 is unserviceable in Paris (CDG). The engineers report that it has a hydraulic leak such that it may require a hydraulic pump change. If so, then they expect the pump change to take two hours. On this advice, the staff at CDG have stopped checking passengers in for Flight 705. (After participants were given time to consider this situation, subsequent information was provided that confirmed the hydraulic pump change and advised that due to inclement weather, the maintenance work would be done in the hangar delaying a possible departure considerably more than initial advice).

Scenario Two required participants to consider strategies and consequences to resolve the delay caused by the unserviceable aircraft. The flight was to operate Paris (CDG) to Pacific (PFC) but was progressively delayed in CDG due to mechanical unserviceabilities, to the extent that the operating crew were eventually unable to complete the flight within their legal duty time. Figure 7.2 provides a graphical depiction of this scenario.

From several alternative actions that could have been considered, the optimum solution provided by the expert panel was to re-route the flight from CDG to Mumbai (BOM) and position another crew from PCF into BOM to enable the flight to continue to PCF with minimal disruption. The inter-rater reliability for classifying participants' comments into the categories of decision considerations in Scenario Two was .73. The inter-rater reliability for identifying comments within the category was .96. One participant did not take part in Scenario Two due to an administrative oversight.

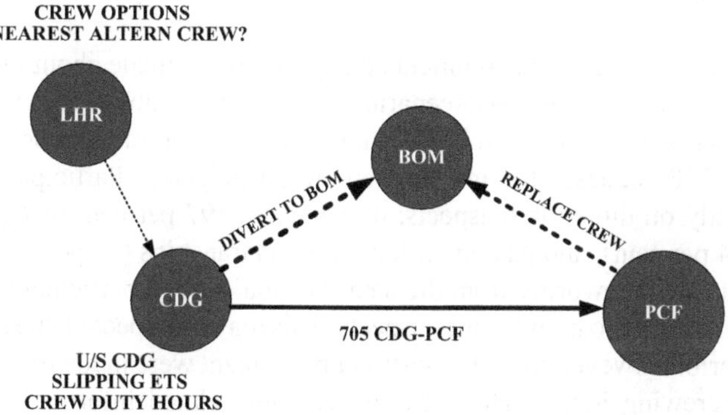

Figure 7.2 Graphical representation of international scenario two. See Appendix A for international utilisation (not to scale)

Revised Situation Awareness

Participants were made aware that each scenario in the international simulations was independent of the other scenarios. Accordingly, during Scenario Two, participants enhanced their situation awareness in the course of the scenario briefing predominantly by requesting maintenance information specific to the unserviceable aircraft.

Information Completeness

Participants' comments emphasised the need for accuracy of information about the situation. This was consistent with the findings in Scenario One. However, participants appeared to place greater importance on checking information because the scenario was more complex and information was required from technical experts. For example:

> The problem is the Paris engineers' information update. Sometimes the information from our own port is second-hand information. I would double check with the Paris staff first to see what the answer is. (502)

> ... it looks like we wouldn't press the button on anything else until we had further word from an engineer. It's a regular happening for someone to be told that [the aircraft is] not going anywhere, and then ten minutes later a magical fix has been found, so we wouldn't take any action at this stage of the game. (205)

> So the question [that] has been raised would also have been to [the] crewing [section] – my expertise not being crewing ... to ensure that they also are looking and giving to me the best information they can on availability of crews. (211)

In contrast to the general comments that participants made about the accuracy of information in the previous scenario, the comments above were specific to the maintenance issues concerning the unserviceable aircraft. In Scenario Two, there were 258 requests for information by participants. Participants focused predominantly on three main aspects: maintenance (97 per cent of participants), crewing (94 per cent), and passenger loadings and tranships (72 per cent). Several observations are noteworthy from the data. The majority of participants requested information relating to maintenance aspects. This may be expected due to the focus of the scenario. However, more requests per participant were made for information relating to crewing issues. The focus on crewing-related information was also evident in the previous scenario. It appears from these results that participants may regard certain fundamental information as necessary for increasing situation awareness irrespective of the situation.

Decision Considerations

Participants made 201 comments relating to decision considerations in Scenario Two. The comments were classified according to elementary (17 per cent), core (58 per cent), and advanced (25 per cent) categories.

Elementary Decision Considerations

Comments were classified as elementary when participants appeared to acknowledge the information provided and began to consider the basic consequences of the scenario. This was similar to the process used in the previous scenario. Although the problem in Scenario Two concerned an unserviceable aircraft, participants were told that the initial information about the problem was reported by the airline's ground handling staff. Despite the comments above relating to checking the reliability of maintenance information, some participants (13 per cent) readily accepted the initial information about the aircraft. For example:

> Ok ... being a port where we don't have any ability to change aircraft, we are basically locked into that delay ... basically the information the engineer has given us is ... what we would stick to ... a two hour delay. (201)

Comments were also classified as elementary if they related to basic considerations such as identifying opportunities to replace the unserviceable aircraft, or transhipping passenger problems. This level of comments was consistent with the elementary comments in Scenario One. For example:

> [I'm] looking to see if there's any other aircraft anywhere around Paris ... There's obviously the London one, but I wouldn't want to amend that at all. They're a full ship anyway. [I] can't do anything about that. (311)

Most participants (75 per cent) identified elementary decision considerations. This is consistent with the result for Scenario One (73 per cent) and suggests that most participants seem to identify various basic level considerations regardless of the situation.

Core Decision Considerations

All participants identified decision considerations at the core level which indicated that they were considering more complex consequences of the maintenance problem than those evident at the elementary level. Many participants (25 per cent) did not indicate elementary considerations. This was similar to the result in Scenario One and suggests that either they did not consider aspects such as changing aircraft or transhipping passengers, or they did not verbalise these aspects. Several comments related to identifying potential constraints such as slot times or curfews. However, the majority of comments at the core level (64 per cent) related to crewing restrictions such as crew sign-on time, duty time limitations, or crew replacement alternatives. These considerations were identified by 94 per cent of participants. For example:

> I need to know where my crews are ... would a one hour delay Paris to Pacific necessarily put them at a point where they exceed the tour of duty? (211)

> For now our main consideration is the crew's flight time limitation. Let's say there is another delay. The crew will not be able to operate to their destination. (405)

> [I'd] go to the tech[nical] crew desk and ask them [if] they have any pilots we can call out ... we would ask the tech crew desk to have a look at the crew that's in Paris. Have they had the minimum rest requirements? What time can ... they report to go? (305)

These comments demonstrate that although participants identified these constraints, they appeared to consider ways to minimise the disruption by keeping within the various constraints. Working within these limitations was similar to the way in which they approached the situation in Scenario One.

Advanced Decision Considerations

Around 97 per cent of comments were classified at the advanced level. In contrast to the core considerations, comments classified as advanced decision considerations indicated that participants considered more complex crewing alternatives. This was evident from the comments that referred to extending the crew duty time or seeking replacement crews from other operating ports. For example:

> Ok, we have a ... problem with the crew hours ... what's the maximum discretion they can use? (500)

> ... can we check whether it's possible to get [a] crew from a neighbouring station, position them to Paris and thereafter operate from Paris to Pacific in one tour [of crew duty time]? (400)

This was a similar finding to the comments at the advanced level of considerations in Scenario One. However, in Scenario Two, more participants (75 per cent) identified advanced decision considerations in relation to crewing aspects. This suggests that not only could participants comprehend the complex crewing situation, but they identified ways to reduce the consequences of the situation.

Comments at the elementary level indicated that participants challenged the reliability of information. However, around half of the participants (53 per cent) at the advanced level needed to re-confirm particular information during the scenario. For example:

> OK. I would like to check again with Maintenance that the time they have given us [is] absolute. (307)

> What I would do is contact our Maintenance Watch people here and ask them to get in touch with the [Maintenance] people at Paris, just to confirm that is the problem [and] that the part is available ... I'd be concerned that the original indication of a potential two hour delay seems to be now more or less an indefinite or an unknown delay and I want to get the potential ... impact of that. (315)

These comments suggest that for participants to consider opportunities to overcome situations, they needed to ensure that the information upon which their decision-making processes relied was sound. Similar to the findings in Scenario One, the classification of participants' comments into the three categories of decision considerations demonstrated substantial variation in the way that participants considered the implications of the scenario. To investigate whether this variation may be influenced by participants' levels of expertise, the study investigated the relationship between expertise and decision considerations.

Expertise

The comments made by novices in Scenario Two demonstrated patterns similar to the comments in Scenario One. While their requests for additional information often related to the different decision considerations they identified, novices appeared to draw upon this variety of information in a haphazard way. For example:

> OK, check with Maintenance. Make sure they are aware of [the mechanical problem with the aircraft] as well. [Regarding] crewing ... what hours, what flight hour limitations have we got? Next we could look at where the passengers are going to once they get [to Pacific]. (311)

> Based on the engineer's decision, you would publish the delay of 0855 on advice that if they can get it ready earlier, we would be looking at going. [Then I'd] contact ... Despatch asking if a faster time can be organised to make up time a) for the passengers and b) for the crew tour of duty. Just trying to think. [I'd] stay in contact with Maintenance Watch to make sure that they keep us in the loop if they fix the part earlier. (213)

In contrast, experts demonstrated a far more methodical approach than novices. They clarified information, discussed the problem in far greater detail than novices, and appeared to know what consequences should be investigated in some order of priority. In some cases, they demonstrated highly proactive methods to manage the situation. For example:

> Right ... so I'd offer my resources to actually try and help him find where [a maintenance part] is, because I'm a hunter ... I'd go to [another airline] Maintenance [department] and put the two of them in touch with each other straight away ... I'd ask at this stage for the local people to start looking at other carriers to see what capacity is around. I'd make sure I speak with the captain as soon as the captain checks in. (216)

> If we have this information and ... there is no [aircraft] you can drop in [to Paris], we get a [telephone] line with the duty engineer. We tell him the constraints and the time we have for the delay. We want him to pass his time of serviceability. If that time is way beyond the crew time, we would prefer to stand [the crew] down ... but if the serviceability is between five and six hours, the crew can't operate the flight. (401)

These comments support the findings in Scenario One indicating considerable differences in the way that novices and experts managed the situation. Table 7.10 investigates this variation further by presenting the relationship between expertise and decision considerations in Scenario Two for novices and experts according to the level of expertise in both industry and OCC.

In Scenario One, the findings suggest that participants focused on a set of core considerations, regardless of their level of expertise. A similar pattern emerged in Scenario Two. Therefore, participants appeared to focus on considerations that minimised disruptions even as the complexity of the situation increased. In Table 7.10 the mean numbers for all categories were higher than the corresponding means for Scenario One which suggests that participants identified more considerations at all levels of decision considerations regardless of their level of expertise.

Notably, novices in both industry and OCC recorded the highest mean numbers in nearly every category of decision considerations. This finding differs from the findings in Scenario One and suggests that in some situations, participants may not necessarily rely on their level of expertise to identify considerations. The results suggest that expertise may enable participants to identify and weight decision considerations in some but not all situations.

Table 7.10 Frequencies and mean numbers of decision considerations comments by novices and experts according to the level of expertise in industry and in OCC. (*N*=32)

Decision considerations	Novices in industry[a] and OCC		Experts in industry and novices in OCC		Experts in industry and OCC	
	n=8		*n*=14		*n*=10	
	f	*M(SE)*[b]	*f*	*M(SE)*	*f*	*M(SE)*
Elementary	10	1.3(.25)	13	.93(.25)	11	1.1(.23)
Core	32	4.0(.27)	48	3.4(.36)	37	3.7(.40)
Advanced	11	1.4(.32)	21	1.5(.20)	18	1.8(.25)

[a] No participants were both novices in industry and experts in OCC.

[b] *M* = the mean number of comments per participant in the category.

SE = estimated standard error of the mean.

Decision-making Styles

The extent to which participants used rational and intuitive decision-making styles was examined in relation to Scenario Two. The analysis of data also examined the extent to which these styles appeared to differ between novices and experts.

Rational Decision-making Style

In Scenario Two, the comments from only one participant were classified as low rational DMS. The comments demonstrated little evidence of a systematic approach in the decision-making process. For example:

> Ok. Well, I'd be looking at either ... no I wouldn't because that would affect ... I was looking [or] going to look at swapping aircraft to the [flight] 700/705 [aircraft], but that would be pointless because it would start causing problems into London. What I'd be looking for is an aircraft that could pick up [the flight] out of Paris, but really I thought they were in Pacific, so delete reference to that altogether. (213)

In contrast, comments from the other participants indicated high rational DMS. For example:

> I'd certainly be keeping the engineers going with the pump change even though we have lost the original crew to crew tour of duty problems. There is still a chance of using the inbound [flight] 706 crew for a 1200 [departure], once they have had their rest. If 1200 UTC is their first time they can operate certainly there is a chance of getting the aircraft out of Paris ... even though that is going to be a four hour delay, we have quite a long extended ground time in Pacific. When the aircraft gets back it is on the ground for about 14 hours it looks like. (201)

It was difficult to determine why more participants in this scenario used a high rational DMS than in Scenario One. A possible reason may be the level of familiarity of the simulations gained from Scenario One. Table 7.11 presents the relationship between participants' use of rational DMS and expertise, according to the level of expertise in both industry and OCC.

The data in Table 7.11 indicate that all participants used a rational decision-making style to a far greater extent than they did in Scenario One. This was evident for both novices and experts. Nearly all participants appeared to evaluate the mechanical problem and identify decision considerations in a very systematic way, despite the increased complexity of the scenario. Although a rational decision-making style may not be suitable in complex situations where circumstances change dramatically, participants in the current scenario appeared to use a high rational DMS. This suggests that situation complexity may not necessarily determine the decision-making style used and participants may still use a systematic approach provided that sufficient time is available.

Table 7.11 Classification of participants' use of rational decision-making style according to their level of expertise in industry and in OCC. (*N*=32)

Rational DMS	Novices in industry[a] and OCC *n*=8		Experts in industry and novices in OCC *n*=14		Experts in industry and OCC *n*=10	
	n	%	*n*	%	*n*	%
Low	1	13	-	-	-	-
High	7	87	14	100	10	100

[a] No participants were both novices in industry and experts in OCC.

Intuitive Decision-making Style

The comments of a few participants (16 per cent) indicated that they appeared to use low intuitive DMS in this scenario. For example:

> I guess now it's just a waiting game for more information from maintenance as to whether there's a [replacement maintenance] part or not. I'd give them a bit of time to sort it out obviously ... probably give them say about 15 minutes to make their phone calls and then I would chase up. (212)

> ... it's basically the same information that we have [already]. Then there is no point hanging around. The sooner we step down (sign off) the crews, the sooner we can depart [after crew rest]. (501)

In contrast, the majority of participants used high intuitive DMS. These participants seemed to assess the situation more comprehensively and seemed to know what might or might not work. For example:

> OK ... hydraulic leak. Now that's ... in my own mind I'm thinking that could be actually quite a problem. Immediately I'd get the rest of ... my group looking at factors. I'd just do a quick brief, tell them what the problem is and seeing if this thing fell over permanently what in terms of long-term effect ... (310)

> ... just the time required to change that part and the other factor ... if the weather is not looking too good, then they could probably sit it out for half an hour, the rain's going to pass on and between times to and from the hangar and back again,

> there's not a lot of opportunity to make up time. So it looks like 1100 is totally unrealistic, so there's really no point in keeping the passengers at the airport for two to three hours then to be told after that the flight's not operating ... (503)

Table 7.12 presents the relationship between participants' use of intuitive DMS and expertise according to the level of expertise in both industry and OCC.

The data in Table 7.12 indicate that more participants, both novices and experts appeared to use an intuitive decision-making style during Scenario Two than they did for Scenario One. These findings suggest that with greater complexity, both novices and experts rely increasingly on their gut feeling for problem solving. However, it appears that participants with industry and OCC expertise may use intuition to a greater extent than novices. In contrast, there was little difference between novices and experts in Scenario One, where the situation was less complex.

Table 7.12 Classification of participants' use of intuitive decision-making style according to their level of expertise in industry and in OCC. (*N*=32)

Intuitive DMS	Novices in industry[a] and OCC *n*=8		Experts in industry and novices in OCC *n*=14		Experts in industry and OCC *n*=10	
	n	%	*n*	%	*n*	%
Low	3	38	2	14	-	-
High	5	62	12	86	10	100

[a] No participants were both novices in industry and experts in OCC.

Generation of Decision Alternatives

In Scenario Two, the main decision alternatives that participants needed to examine related to crewing limitations. Many participants (66 per cent) identified ways that enabled the flight to operate to a delayed schedule by extending the crew duty time. However, information provided during the scenario announced further delays to the flight, such that the crew could not operate within this extended duty time. Therefore, participants were required to consider other decision alternatives. About half of the participants (53 per cent) indicated that one alternative was to wait for the crew to take crew rest and operate the flight the next day. Of these participants, 35 per cent were novices in OCC and industry, 35 per cent were novices in OCCs, but considered expert in industry, and 30 per cent were considered expert in both industry and OCC. Other participants (34 per cent) identified an alternative which could enable the flight to depart and operate via another port that bypassed the constraints of the crew duty time. Only 18 per cent of these participants were novices in OCCs. The other participants were expert in either industry or OCCs. The findings support the findings in Scenario One, suggesting that participants with greater expertise appear to identify more creative decision alternatives to overcome limitations.

Scenario Three

The following initial briefing was provided to participants at the start of Scenario Three:

> The time is 0100. There is an alternate[1] on Pacific due to a typhoon warning level 5 (medium intensity). The typhoon is stationary at present 140 nautical miles north east of Pacific. (Information provided to participants later in the scenario indicated changing conditions including intensification of the typhoon, movement and speed toward Pacific, and expected impact and damage at the airport.

Scenario Three required participants to consider the operational impact of the approaching typhoon on flights operating into and out of the airline's main airport of Pacific. Figure 7.3 provides a graphical depiction of this scenario.

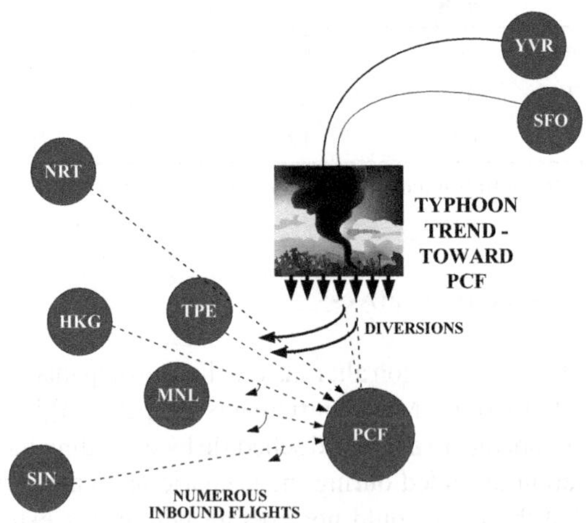

Figure 7.3 Graphical representation of international scenario three. See Appendix A for international utilisation (not to scale)

1 An *alternate* requirement at an airport indicates a legal obligation for an aircraft to carry sufficient fuel to enable it to divert away from its intended destination to an alternative, suitable airport. Thus, an aircraft destined for Pacific must carry sufficient fuel for it to divert safely to, say, Hong Kong in the event it cannot land at Pacific.

The expert panel did not provide an optimum solution for this scenario due to the high complexity of the scenario and the numerous alternatives that participants could produce. However, the panel indicated several considerations that participants should take into account during the decision-making process, such as questioning the progress of a typhoon, identifying aircraft at risk, and holding aircraft at their originating airport until weather conditions improved. The inter-rater reliability for classifying participants' comments into the categories of decision considerations in Scenario Three was .75. The inter-rater reliability for identifying comments within each category was .89.

Revised Situation Awareness

In Scenario Three, participants enhanced their situation awareness in the course of the scenario briefing and by requesting information specific to operations likely to be affected by the weather conditions at Pacific.

Information Completeness

When Scenario Three was introduced, nearly all participants (97 per cent) accepted the information as it was given and began to assess the consequences of the weather situation immediately. This approach differed from the previous two scenarios where several participants initially confirmed the accuracy and source of the information. However, several participants then clarified certain aspects of the weather situation. For example:

> We need to get an estimate of the route of the cyclone. (205)
>
> At this stage I don't know if there's really very much we can do, except keep in touch with the weather bureau and find out if [the typhoon] starts to move. (214)
>
> If I hadn't known [the typhoon] was stationary, I'd be ... looking at the direction and getting a meteorological forecast from the operations despatch department [to] try and ascertain where it's going to go. (213)

In Scenario Three, there were 183 requests for information by participants. Participants focused predominantly on three main aspects: crewing (73 per cent of participants), weather (85 per cent), and flight planning and despatch (70 per cent). The focus of comments in relation to weather conditions could be expected in view of the type of scenario. In common with the previous scenarios, participants emphasised crew-related issues. It appears that participants regarded crewing issues in particular as central to the situation regardless of the scenario. In the previous scenarios most participants requested information about passenger loadings on flight. However, in the current scenario very few participants (18 per cent) asked about loadings. This suggests that information that may be regarded as important for situation awareness in some situations may not be deemed as crucial in another situation. Therefore, participants may have insufficient awareness to manage the disruption.

Decision Considerations

Participants made 236 comments relating to decision considerations in Scenario Three. The comments were classified according to elementary (18 per cent), core (36 per cent), and advanced (46 per cent) categories.

Elementary Decision Considerations

Comments classified as elementary indicated that participants appeared to identify basic level considerations such as aircraft patterns directly affected by the weather situation as well as passenger tranship problems at Pacific. A few participants (12 per cent) commented specifically about the current strength or position of the typhoon. Most participants (82 per cent) considered operational disruptions likely to occur at Pacific by comparing the estimated arrival time of the typhoon at Pacific with the arrival and departure times of flights scheduled through Pacific. For example:

> [I look at] how many flights we've got en-route to Pacific; how many flights have we got due to depart from a port and heading to Pacific; how many flights have we got leaving from Pacific. That would be my first thought. (212)

Some participants (18 per cent) indicated that until the typhoon developed further, they would not take any specific action. For example:

> OK, then. I guess at this stage, it would just be op[eration]s normal until we get information to suggest that the typhoon is heading our direction. (212)

Core Decision Considerations

Comments classified at the core level demonstrated that participants had better comprehension of the consequences of the typhoon, than shown by comments at the elementary level. This was similar to their approach in the previous scenarios. In contrast to basic considerations such as identifying flights disrupted during the typhoon, core considerations demonstrated concern for managing the disruption. This was evident from comments relating to specific flights. For example, participants considered the maximum amount of time that aircraft enroute to

Pacific could circle before either having to land at Pacific or divert to another airport.

> So we'd find out first of all what fuel [aircraft registration] LHE and [aircraft registration] LHF have. We'd let them come on [to Pacific] as they have got the alternate (an airport to which the aircraft may safely divert) covered. It is also a reasonably ideal situation that they have got fuel to actually make one attempt [at landing], where if you didn't have fuel to make the attempt, you'd come as far as you could [while] updating the [weather] forecast ... (205)

> So we go through all these flights that are due at that time of the day and identify what the alternates are. Basically from experience you would know the alternates and how many aircraft they can take. (503)

Participants also identified critical fuel conditions on very long-range flights as their flight-paths passed through the typhoon. For example:

> So, at this stage, I'd be ... talking to [the] flight planners now and saying, look – if this thing moves to the south we can still operate. Pacific is still operating [but] how will this affect the Vancouver to Pacific and San Francisco to Pacific [flights]? (310)

These comments demonstrate that participants focused on more complex consequences of the disruption beyond identifying specific disrupted flights; yet it appears that they managed the disruption within the constraints they had identified. For example, while they might have identified suitable alternate airports, they may not have considered the consequences of diverting a number of aircraft to one port. The results suggest that while some participants may identify limitations of a situation, they may only take into account consequences of a situation that correspond with these limitations. Therefore, core level considerations may not be sufficient to enable participants to manage disruptions to gain optimal results in highly complex situations.

Advanced Decision Considerations

Many participants (33 per cent) looked for specific ways to lessen the consequences of the weather situation on flights operating through Pacific by seeking further information about the projected path of the typhoon and expected conditions on its arrival at Pacific. This information enabled them to consider alternative actions such as increasing fuel loads on aircraft about to depart for Pacific, changing the departure times of flights to avoid the worst weather conditions, and either securing aircraft on the ground or evacuating them from the area. For example:

> Ok. I would look at holding ... delaying the flight ... [then] ringing our outstations [and] advising them of what's happened, [then] try[ing to] get hold of the [captain] and looking at possibly delaying that flight to depart at a later [time] ... until we know exactly what's happening with the typhoon. (305)

> First of all I would be contacting Taipei and saying we want additional fuel on [aircraft that are] there ... and I'd like them to check with us prior to closing the [aircraft] door [as to] whether we're going to release it (let the aircraft depart) or not. (308)

Several participants (52 per cent) considered likely consequences of their plans such as congestion at airports should a number of aircraft divert to the same airport. For example:

> [I'd] call first Manila ... and then Taipei. Let's get them alerted, check the weather and the situation there ... [for example, to see if] there [are] plenty of parking spots ... then I can work out a plan based on their directions and how many [aircraft diversions] they can accept and how many they can handle. (508)

It appears that participants who identified advanced decision considerations were more likely than others to manage disruptions effectively. To investigate the way in which these considerations may be influenced by participants' levels of expertise, the study investigated the relationship between expertise and decision considerations.

Expertise

In Scenario Three, novices appeared confused with the initial information provided about the typhoon. Many novices requested information to be repeated or clarified to enable them to determine the position, strength, direction, and speed of the typhoon. For example:

> [The typhoon is] moving in which direction? (201)

> So the typhoon is ... what did you say about [it]? And they can't give us any forecast as to whether it's going to ... (212)

> Can you start again? Sorry? What is the time? (400)

In contrast, experts appeared to comprehend the information more readily. This was likely to be due to their previous experience or knowledge of similar events. For example:

> The weather pattern is that [the typhoon] is stationary there. That would bring a lot of weather over Pacific. There is a lot of moisture circulating through the area. Conditions would be fairly minimal. Alternates or disruptions are obviously significant. I would be looking straight away at [the conditions at] Taipei. (203)

> Well, depending on the seriousness and the strength of the storm, you may want to actually get your aircraft out of the port ... you have ... usually got so many aircraft there that your chances of being able to move them out to somewhere out of the range of the storm is probably not really practical but what you can do ... certainly with your international aircraft [is] try and plan for an alternate destination if, when you get there, the place is closed. (205)

Table 7.13 presents the relationship between expertise and decision considerations for novices and experts according to their level of expertise in both industry and OCC.

The findings from the first two scenarios indicated that participants appeared to focus on core level considerations, regardless of their level of expertise. In Scenario Three, this was less evident. Rather, participants identified more advanced considerations. This suggests that in complex situations, participants may focus on ways to overcome a disruption, rather than manage the disruption within constraints. This approach was apparent for both novices and experts. This finding is important as it suggests that participants do not need to be experts necessarily to demonstrate their ability to anticipate and overcome situations. The findings also suggest that experts may not necessarily perform better than novices in a situation. Thus, the results are consistent with previous research by Hershey

Table 7.13 Frequencies and mean numbers of decision considerations comments by novices and experts according to the level of expertise in industry and in OCC. (N=33)

Decision considerations	Novices in industry[a] and OCC		Experts in industry and novices in OCC		Experts in industry and OCC	
	n=8		*n*=15		*n*=10	
	f	*M(SE)*[b]	*f*	*M(SE)*	*f*	*M(SE)*
Elementary	10	1.3(31)	20	1.3(.29)	13	1.3(.21)
Core	16	2.0(.50)	36	2.4(.29)	33	3.3(.50)
Advanced	29	3.6(.53)	43	2.9(.43)	36	3.6(.52)

[a] No participants were both novices in industry and experts in OCC.

[b] M = the mean number of comments per participant in the category.

SE = estimated standard error of the mean.

and Walsh (2000) who found that novices who had undergone limited training outperformed experts in an auditing task.

Decision-making Styles

The extent to which novices and experts used rational and intuitive decision-making styles was examined in relation to Scenario Three.

Rational Decision-making style

A few participants (15 per cent) made comments that were classified as low rational DMS. The comments showed that participants considered several aspects of the scenario. However, the comments often appeared unconnected, revealing little evidence of any systematic process. In this way, the comments showed similar patterns to comments relating to low rational DMS in the previous two scenarios. For example:

> I would be looking at landings in this area. Is the airport safe? I would be concerned for the fuel policy. I'm ... looking at destinations with a curfew. (506)

> The conditions may get worse and worse, or better. Then we'd have to do something. Call ... Manila and then Taipei. Let's get them alerted, check the weather and the situation there. [Are] there plenty of parking spots? [I'd also call] Engineering Support. Are they OK to accept more than diversion flights? (507)

Comments classified as high rational DMS indicate that some participants used a highly systematic process to manage the weather situation in Scenario Three. For example:

> I have been told no trend type forecast, so I am not looking at anything which may necessarily be forecast to go towards, or track towards a particular port as such ... The obvious thought ... is [that] we have a typhoon level five, it's a medium typhoon [so] we obviously have some lousy weather. It will by definition still have an effect on flight planning and on en-route weather. I would therefore be [speaking] with Flight Planning getting the best information ahead as to what tracks we are using. (211)

> OK well that's fine ... well I'd just continue on the track at the moment. I'd advise the [captains] of what we're looking at, [that] if [the typhoon] comes down and starts affecting the track, [the plan] would be to divert around it and if they have sufficient fuel, to still keep that one approach. If the weather is still ... pretty bad, [we'd look at] going to Manila and to Taipei. [Then] refuel and

> be prepared to come immediately back again, presuming by that stage that the weather's moving through quite quickly (310)

The differences in these comments indicate the variation in the way in which participants used a rational DMS. Table 7.14 presents the relationship between participants' use of rational DMS and expertise according to the level of expertise in both industry and OCC.

The findings from Table 7.14 were consistent with the findings in Scenario Two. Although Scenario Three was more complex than Scenario Two, participants maintained high rational DMS to manage the situation. This supports the findings in Scenario Two suggesting that both novices and experts appear to rely more on rational decision-making style in highly complex situations.

Table 7.14 Classification of participants' use of rational decision-making style according to their level of expertise in industry and in OCC. (*N*=33)

Rational DMS	Novices in industry[a] and OCC *n*=8		Experts in industry and novices in OCC *n*=15		Experts in industry and OCC *n*=10	
	n	%	*n*	%	*n*	%
Low	2	25	2	13	1	10
High	6	75	13	87	9	90

[a] No participants were both novices in industry and experts in OCC.

Intuitive Decision-making Style

In Scenario Three, comments that were classified as low intuitive DMS indicated that while participants considered aspects of the typhoon likely to disrupt flights, there was little evidence to suggest that participants had a strong sense of what might work or not work. For example:

> [The typhoon's] 140 miles out. It can change direction at any time and it could affect all the other ports. I will make the flights go still on time, but I can tell them to get additional fuel. (507)

> We would basically be looking at keeping a monitor on that, on ... what this typhoon's going to be doing. Obviously we would be in touch with our ... flight planners and look at what routes [the aircraft] are going to be taking to ... go around ... this typhoon. (302)

Only 15 per cent of participants' comments were classified as low intuitive DMS. In contrast, most comments in this scenario indicated that participants appeared to use a high intuitive DMS. Many comments related to the participants' sense for anticipating the consequences of the typhoon. For example:

> ... they said this thing's moving south west at 40 knots. So I'd be diverting to a port that was not in line with the forecast track of [the cyclone]. My issue there would be, even though I'm given these rubbery figures for [the cyclone] hitting the airport, if it comes earlier, the damage to the infrastructure of the airport could be ... [to] the navigation aids so there [are] no [navigation aids] for [the Vancouver flight] to make an approach anyway. (217)

> ... there's a very high possibility of them coming in at the same time as the typhoon will hit us and therefore a strong possibility of them not being able to land at Pacific and having to divert somewhere else and we don't want that. (400)

Table 7.15 presents the relationship between participants' use of intuitive DMS and expertise according to the level of expertise in both industry and OCC.

The findings in Table 7.15 indicate that both novices and experts appeared to use high intuitive DMS. The results also support the findings of the previous scenario indicating that experts appeared to use intuition more extensively than novices in complex situations.

Table 7.15 Classification of participants' use of intuitive decision-making style according to their level of expertise in industry and in OCC. (*N*=33)

	Novices in industry[a] and OCC		Experts in industry and novices in OCC		Experts in industry and OCC	
Intuitive DMS	*n*=8		*n*=15		*n*=10	
	n	%	*n*	%	*n*	%
Low	2	25	2	13	1	10
High	6	75	13	87	9	90

[a] No participants were both novices in industry and experts in OCC.

Generation of Decision Alternatives

In Scenario Three, participants were required to consider the operational impact of an approaching typhoon for which the location, size, direction, and speed of the typhoon had been provided. Although many participants (33 per cent) commented that they would let all flights operate according to their schedule, several participants (42 per cent) began to consider contingency plans. As further information about the typhoon was provided, participants identified specific alternatives such as nominating available airports to which aircraft could be diverted (24 per cent) and delaying specific departures to avoid arriving at Pacific during the typhoon (39 per cent). As the scenario did not extend to the impact of the typhoon at Pacific, the important findings for the study relate to the extent to which participants identified decision alternatives. However, it was evident from the comments that participants with industry and OCC expertise appeared to prepare more comprehensively for the disruption than novices.

Summary of Trends Across the International Scenarios

A number of patterns emerged from studying decision-making processes in the three scenarios of the international simulations. Participants accumulated situation awareness by seeking information specific to the scenario. In complex scenarios, they emphasised the importance of receiving accurate and reliably sourced information and obtaining further information as circumstances changed. Participants also focused on both elementary and advanced decision considerations. This suggests that in complex situations, it appears critical to take into account considerations of a basic nature such as tranships and weather conditions, as well as considerations that might enable decision-makers to formulate creative solutions to problems such as re-routing flights to ensure their ability to operate. It was also evident that in highly complex situations, both novices and experts focused on advanced decision considerations. This suggests that participants may not have relied necessarily on their level of expertise to provide creative solutions to problems. Despite evidence in the literature suggesting that a rational decision-making style may be inadequate in complex, dynamic situations, the use of a high rational decision-making style was evident in complex scenarios in this study. The findings indicate that participants also used a high intuitive decision-making style, in particular in highly complex scenarios. Thus, with respect to the international simulations it appears that decision-making relies on both rational and intuitive decision-making styles in complex situations.

Summary

This chapter presented the findings and analysis in relation to decision-making processes of the 33 participants during the international simulations. Participants' comments were classified according to the degree of complexity into three levels: elementary, core, and advanced situation awareness. The classifications indicated that while some information provided a superficial level of situation awareness, other information helped participants identify areas of potential disruption. This method of classification builds on previous work by Endsley (1995a) by suggesting the existence of a sub-level of elements within Level 1 (perception of elements) of her situation awareness model.

The relationships among participants' levels of experience, expertise, and situation awareness were examined. The results indicated that participants acquired initial situation awareness which was enhanced by scenario-specific information. Thus, it was evident that acquisition of situation awareness was a cumulative process. While the importance of domain experience for acquiring situation awareness has been referred to previously, the findings in the international simulations emphasise that participants' levels of industry experience are also important. Participants classified as industry and OCC experts focused on advanced levels of information which enabled them to build more comprehensive situation awareness than novices.

The study took into account the decision considerations made by participants. These considerations were classified according to the degree of complexity into three levels: elementary, core, and advanced. While some participants recognised superficial considerations, other participants considered alternative ways to overcome or avoid potential disruptions. The decision considerations step is not explicit in previous models of decision-making. However, it is evident from participants' comments in the international simulations that this step is important to the decision-making process.

The study examined the relationships among expertise, decision-making styles, and decision considerations. The results indicate that novices approached the decision-making process in a sporadic manner, often focusing on superficial aspects of situations. In contrast, experts developed a more comprehensive approach, enabling them to recognise and work beyond problem limitations. The increased focus on advanced decision considerations by both novices and experts in complex situations was an unexpected result in the international simulations as it was presumed that novices lacked sufficient experience to consider these aspects. The extent to which participants used a rational decision-making style and an intuitive decision-making style was examined and the findings indicate that participants relied on both styles regardless of the complexity of the international simulations. This was an unexpected result and is inconsistent with previous arguments that a rational decision-making style may be inadequate in complex, dynamic environments. Thus, the findings are in contrast to the work by researchers such as Beach and Mitchell (1978) and Hammond (1993) who suggested that decision-making lies within a continuum from analytical to intuitive extremes. In contrast, the findings suggest that decision-making uses a combination of both intuitive and rational approaches.

Chapter Eight presents the findings and analysis in relation to decision-making processes of the 19 participants during the domestic simulations. Comparisons between the data regarding the international simulations in the current chapter and the data regarding the domestic simulations are made throughout Chapter Eight.

Chapter 8

The Domestic Simulations

Introduction

This chapter presents the findings and analysis in relation to decision-making processes by the 19 participants in the domestic simulations. To identify differences between international and domestic decision-making processes, comparisons between the findings for the international and domestic simulations are made throughout this chapter. Thus, the structure of this chapter replicates the previous chapter.

Background to Domestic Simulations

It is vital in scheduling domestic operations to optimise the schedule in order to maximise resources such as aircraft and crews. As domestic schedules are usually contained within one calendar day this optimisation usually results in very high utilisation of aircraft with minimal time between flights. As a result, disruptions to domestic schedules tend to be more extensive than disruptions to international schedules and recovery far reaching and more complex. Hence decision-making often requires some combination of flight delays, cancellations, aircraft changes with numerous operational and commercial consequences. This is considerably different from international schedules where the longer flight stages – often beyond one calendar day, more isolated disruptions and recoveries enable decision-making over long time-frames. Given the complexity and severe consequences of disruption to domestic schedules, timely decision-making is even more critical to schedule integrity.

Conducting the Domestic Simulations

Similar to the international simulations, the domestic simulations consisted of two stages: an initial stage where participants familiarised themselves with the domestic flight schedules and a second stage requiring participants to assess the situations in three individual scenarios and explain their decision-making processes.

Initial Situation Awareness

Participants familiarised themselves with the domestic flight display prior to the three scenarios. No operational disruptions were occurring during this initial familiarisation process. Participants' comments indicated that they observed similar aspects of the flight display as participants in the international simulations, namely: the flight schedules, gaps in the aircraft utilisation, maintenance requirements, passenger loadings, and weather. Many participants (68 per cent) indicated that they used the familiarisation stage to prepare for operational problems and look for weaknesses in the utilisation as a means of overcoming these problems. For example:

> All I'm doing here is highlighting some of the areas that can be utilised if there's any sort of disruption. [I'm looking at] mainly the gaps ... the training [flight], and the ground time of the aircraft in the various ports. (103)

> I would have a quick look ... at the main ports anyway ... just on the assumption that something may go wrong on the first flight of the day and basically [I] just do a quick mental shuffle as to how far I could clear an aircraft type out in any given port. (210)

There appeared to be considerable emphasis on identifying sources of potential disruption, particularly with regard to maintenance aspects. This was evident from a very high concentration of comments (95 per cent of participants) relating to maintenance requirements, limitations, and concerns for potential delays. For example:

> Maybe there might be an aircraft coming out of maintenance. I find that's very important because normally if it is coming out of maintenance, chances are that [it] may not [be in time for] its departure. (202)

> I'm also interested in the maintenance hours left to run to check time, which gives us an idea of what we can do with an aircraft beyond today. (303)

Most participants (74 per cent) identified gaps between flights. For example:

> The thing I would be looking at would be sit-over aircraft ... meaning that this [aircraft] NBM would be sitting in Sydney for, I would say, three hours there ... and there's one here, NBO that would be sitting in Melbourne for, say, two hours. (314)

The existence of gaps or long ground times built into a utilisation during schedule construction can be critical for disruption management in airline OCCs as they provide ways to absorb delays and other disruptions. Although many participants (61 per cent) identified gaps in the international simulations, it was evident from their comments that the identification of gaps appeared to be more critical in the domestic simulations than in the international simulations.

Information Completeness

During the familiarisation stage, participants indicated that they needed additional information to supplement their initial observations. For example:

> [I'd] talk to Coolangatta later [to] make sure they can handle this aeroplane with a U/S APU (unserviceable auxiliary power unit) and make sure the ground power unit is serviceable ... I'd chat to Brisbane, Sydney, and Melbourne; the major ports and say 'have you guys got any problems as yet, that you haven't told me?' (207)

> ... I would be asking Maintenance [as to] which aircraft are running late from overnight schedules, which ones are going to make [departure time], and if they know of any problems that can add at this time ... The protocol is that [at] five o'clock when I walk in, you say what's going to make it and what's not. It's a two way street. They can ring us, or we can ring them. (208)

> If all is OK from a crewing point of view everything is running well. Therefore, crews are connecting and if I don't seem to have any delays ... I am not going to run into any immediate problems. So other than finding out what is running late my next step is to find out what my crewing problems are individually. (202)

These comments suggest that observation alone was insufficient for gaining situation awareness and further information was needed. In the international simulations, most of the enquiries for information during the familiarisation stage related to maintenance, passenger loadings and tranships (connecting passengers), regulatory constraints, crewing, and weather. In the domestic simulations, most enquiries related to maintenance (58 per cent of participants), weather (32 per cent), crewing (26 per cent) and passenger loadings and tranships (16 per cent). The findings indicate a degree of commonality between the simulations. However, given the complexity of domestic schedules, the emphasis on maintenance and crewing considerations was not surprising and demonstrates the extent to which participants regard these aspects as critical for gaining situation awareness.

Classification of Situation Awareness Comments

For consistency with the analysis in the previous chapter, comments were weighted and classified according to the degree of complexity into the same three sub-levels of situation awareness: elementary, core, and advanced. The inter-rater reliability for classifying participants' comments into the categories of situation awareness was .76. The inter-rater reliability for identifying comments within the sub-categories was .70. There were a total of 112 comments for situation awareness which were classified as elementary (54 per cent), core (32 per cent), and advanced (13 per cent).

Elementary Situation Awareness

Comments were classified as elementary when participants' observations appeared to be based on information readily obtained from the display. It was evident from their comments that participants appeared to gain only a fundamental level of situation awareness. At the elementary level, comments in relation to the domestic simulations were similar to comments made in the international simulations. For example, participants in both simulations indicated that they undertook a general overview of the utilisation and observed passenger loadings on flights. However, in the domestic simulation nearly all participants (95 per cent) emphasised an awareness of maintenance-related information such as limitations, requirements, and potential problems. For example:

> I am looking for ... the number of cycles (take-offs and landings) each particular aircraft is doing, the maintenance at the end of the day, and whether an aircraft has to be in that port for maintenance. (101)

> The next thing I would go looking at is ... what maintenance is required that night. I have got maintenance tonight on ... seven aeroplanes and my endeavour would be to keep that maintenance as planned. Then I would be looking at unserviceabilities on aeroplanes [such as aircraft registration] NBN, [which has] an unserviceable auxiliary power unit. (207)

These comments highlight the importance placed on maintenance in domestic operations. This emphasis may be due to participants' awareness of an airline's requirement to conduct maintenance servicing generally at night time and within time constraints between late evening and early morning schedules. The comments also indicate that an awareness of maintenance concerns appears to be central to gaining a basic overview of the domestic operations.

Comments at the elementary level of situation awareness were classified further into seven sub-categories, according to the focus of the comments. Table 8.1 presents the frequency and percentage frequency of comments classified within the elementary level of situation awareness.

The data in Table 8.1 demonstrate the diversity of sub-categories considered as fundamental to participants' situation awareness. It was evident that the sub-categories identified at the elementary level were far more extensive in the domestic simulations than the international simulations. This suggests that gaining a basic level of awareness may be more complex in domestic operations. All participants identified elementary level sub-categories and nearly all of them (95 per cent) identified core level sub-categories. Around two-thirds of participants (63 per cent) identified sub-categories at all levels of situation awareness. The high concentration of responses at the elementary level suggests that participants needed an appreciation of several aspects of the flight display just to gain a fundamental level of situation awareness.

Table 8.1 Classification of comments within the elementary level of situation awareness. (*N* = 19)

Sub-categories identified by participants	f	%
General overview	5	8
Aircraft schedule or utilisation	12	20
Short turnarounds	4	7
Passenger loadings	11	18
Aircraft ex maintenance	7	11
Maintenance release hours	9	15
Current/planned maintenance requirements	13	21
Total number of comments	61	100%

Core Situation Awareness

Comments classified as core indicated that participants also identified more comprehensive information, in particular information that might assist them in anticipating disruptions. The most noteworthy observation of the responses at the core level was the identification of gaps in the utilisation. Although schedules are constructed to maximise the aircraft utilisation, the gaps between flights provide ways for controllers to substitute aircraft as a means of minimising disruptions. These gaps appear to be very important in domestic operations due to the high utilisation of aircraft and short ground times between flights. For example:

> Well I guess I am looking at the gaps, which to me gives me some indication of where I have got some fat in the system if you like. (204)

> NBM (aircraft registration) has a three hour gap between arriving [in] Sydney at 1330 and departing at 1630. There are also a couple of smaller gaps there for

> other aircraft, but NBM is the one that stands out the most. This is significant because ... if that [other] aircraft were to break down in Adelaide, a logical step [would be] to slide that [flight] up there and start swapping things around. (102)

Nearly half of the participants (42 per cent) commented that they looked for signs of potential disruption, such as crewing or weather problems. For example:

> My other immediate reaction on handover would be to have a look at the weather and see what's actually going on. I'd [also] ask what crew restrictions there are, if any, [and] any routines on very high crew hours. That would be [one of] my initial questions on handover. (314)

Many participants (32 per cent) also identified specific operations such as training flights. These flights are scheduled to provide specific pilot training and do not carry passengers. For example:

> [I see we've] got some training (indicated by a training flight on the display) here. [Given] any unserviceability throughout the day, the very first thing to get cancelled would be training. Running RPT (Regular public transport) services takes preference over any sort of training. We try if at all possible ... to keep training intact if we can, but at the end of the day, you've got aircraft and you've got to make money. (103)

> ... you can start looking through [the day] and say ... that you may have to do without the training for a start and basically check out how important ... that training [is] in Melbourne. (304)

The participants suggested that a training flight represents an opportunity for postponement or cancellation during a disruption as it could enable controllers to use the aircraft for passenger flights instead. Consistent with comments classified at the core level in the international simulations, the comments here indicated that participants looked for potential threats that might disrupt the operation as well as weaknesses that could be exploited to help resolve such problems. This demonstrates that participants sought a degree of situation awareness beyond the elementary level. Comments at the core level of situation awareness were classified further into seven sub-categories, according to the focus of the comments. Table 8.2 presents the frequency and percentage frequency of comments classified within the core level of situation awareness.

The data in Table 8.2 indicate that participants appeared to concentrate on aspects of the flight display likely to cause disruptions, or those that could provide them with opportunities to limit or recover from disruptions. This suggests that participants were not satisfied to use the familiarisation stage only as a means of gaining basic information. Rather, they seemed to use this stage to seek a better state of preparedness for disruptions.

Table 8.2 Classification of comments within the core level of situation awareness. (*N*=33)

Sub-categories identified by participants	f	%
Aircraft schedules and patterns	28	32
Gaps between flights	20	23
Maintenance requirements or problems	26	30
Crew connections and duty limitations	12	14
Minimal passenger tranship times	1	1
Total number of comments	87	100%

Advanced Situation Awareness

In addition to core level comments that highlighted the identification of gaps in the utilisation, comments at the advanced level indicated that participants combined these observations with other information such as passenger loadings on flights. In this way, a far more comprehensive level of situation awareness could be gained. For example:

> Also [I have] just a quick glance at the excess ground time in any of the utilisation again, just as a means of absorbing delays if they come to anything ... . Actually I would have a quick look at ... the main ports anyway; Melbourne, Sydney, and Brisbane, just on the assumption that something may go wrong on the first flight of the day and basically just do a quick mental rearrange as to how far I could clear an aircraft type out in any given port. (210)

> Well, if you have a problem with a wide-body [aircraft], you might be able to substitute a narrow-body [aircraft], or combine flights going to one port, if the loadings fit. (303)

The level of preparedness was also evident in relation to potential maintenance problems. For example:

> Then what I would be doing is cross checking to make sure that what aircraft we have on what route is capable of doing that [in relation to maintenance restrictions]. (208)

> So with the APU (auxiliary power unit), I'd specifically look at that [aircraft registration] NBN with the APU unserviceable [and] what it's flying. I might

> have a chat [with] Coolangatta [airport staff] later to make sure that they can handle this aeroplane [and] make sure that the ground power unit is serviceable. Then it's going to have no effect on the rest of the day's flying. (207)

Comments such as these suggest that participants who identified advanced sub-categories of situation awareness sought a state of readiness for likely disruptions and considered possible solutions that overcame the constraints of a situation. Comments at the advanced level of situation awareness were classified further into five sub-categories according to the focus of the comments. Table 8.3 presents the frequency and percentage frequency of comments classified within the advanced level of situation awareness.

The data in Table 8.3 indicate that participants looked for more sophisticated ways to overcome potential problems by reducing the consequences of disruptions, or identifying constraints that could be varied to enable a disruption to be minimised. For example, further to identifying short turnarounds or restrictive maintenance requirements, it was evident that participants looked for opportunities to change aircraft patterns and alternative ways in which maintenance requirements could be satisfied.

Table 8.3 Classification of comments within the advanced level of situation awareness. (*N*=33)

Sub-categories identified by participants	*f*	%
Critical operations	8	20
Potential for change/flexibility	8	20
Potential weather problems	10	26
Market conditions	1	3
Regulatory constraints	12	31
Total number of comments	39	100%

Experience

To investigate whether the variations in levels of situation awareness were influenced by participants' experience, the study investigated the relationship between situation awareness and experience. In the domestic simulations, 63 per cent of participants had aviation industry experience prior to their current role in the OCC. This experience included airport despatch, crew scheduling, fleet scheduling, air traffic control, and experience gained as a pilot or flight engineer. To investigate the relationship between situation awareness and experience in the domestic simulations, experience was examined according to the number of years in which participants had been in the industry and the number of years they had been in the OCC. The frequencies of participants' comments were analysed according to years of experience by comparing the mean numbers of comments, where the mean (*M*) was determined as the number of comments in a category divided by the number of participants in the category who made the comments. The frequencies and mean numbers of comments per participant for situation awareness were classified according to experience in industry (Table 8.4) and experience in OCC (Table 8.5).

The data in Table 8.4 indicate a high concentration of responses classified at the elementary level of situation awareness regardless of the extent of industry experience. In contrast, most responses in the international simulations were classified at a core level of situation awareness. Comparisons of sub-categories of participants' comments suggest that to provide a fundamental level of awareness in domestic operations, controllers may regard aspects such as fleet utilisation, maintenance requirements, and flight loadings as critical. Thus, industry experience in domestic OCCs may enable controllers to identify elementary aspects readily. The relationship between situation awareness and OCC experience was also considered. Table 8.5 examines this relationship according to the number of years participants had been in the OCC.

Table 8.4 Frequencies and mean numbers of comments for situation awareness according to years of experience in the airline industry. (*N*=33)

	Years of experience in the airline industry							
Situation awareness	**1–5[a] n=2**		**6–10 n=6**		**11–15 n=6**		**>15 n=19**	
	f	*M(SE)*[b]	*f*	*M(SE)*	*f*	*M(SE)*	*f*	*M(SE)*
Elementary	3	1.5(.25)	5	.83(.14)	6	1.0(.22)	17	.89(.19)
Core	5	2.5(.25)	14	2.3(.18)	15	2.5(.30)	53	2.8(.25)
Advanced	-	-	7	1.2(.35)	10	1.7(.58)	22	1.2(.32)

[a] There were no participants with less than one year of industry experience.

[b] *M* = the mean number of comments per participant in the category.

SE = estimated standard error of the mean.

The data in Table 8.5 indicate a high concentration of responses classified at the elementary level of situation awareness regardless of the extent of OCC experience. This result is consistent with the findings in Table 8.4 and suggests that OCC experience as well as industry experience in domestic OCCs may enable controllers to identify elementary aspects readily. Similar to the findings for industry experience in Table 8.4, the data in Table 8.5 indicate that there was a high concentration of comments at the elementary level regardless of years of OCC experience. This contrasts with the findings in the international simulations in which the comments were focused on core level sub-categories and suggests that participants in the domestic simulations regarded information at the elementary level to be considerably more fundamental for gaining situation awareness than information at the core level.

In summary, the results are consistent with the findings in the international simulations in terms of explaining how participants' levels of experience may influence the level of situation awareness gained. The highest mean numbers of comments in the domestic simulations were for participants with greater than 15 years' industry experience and greater than ten years' OCC experience. This suggests that as controllers' experience levels increase, industry experience may be more significant for gaining situation awareness than experience in a particular domain. In addition, the emphasis on items classified within the elementary level of situation awareness in the domestic simulations suggests that participants tend to regard these considerations as being fundamental to gaining a basic level of awareness.

Table 8.5 Frequencies and mean numbers of situation awareness comments according to years of experience in OCC. (*N*=33)

	Years of experience in OCC									
Situation	**<1 year**		**1–5**		**6–10**		**11–15**		**>15**	
awareness	***n*=5**		***n*=8**		***n*=10**		***n*=2**		***n*=8**	
	f	*M(SE)*[a]	*f*	*M(SE)*	*f*	*M(SE)*	*f*	*M(SE)*	*f*	*M(SE)*
Elementary	7	1.4(.51)	6	.75(.25)	10	1.0(.15)	1	.50(.50)	7	.88(.23)
Core	12	2.4(.24)	21	2.6(.32)	27	2.7(.26)	5	2.5(.50)	22	2.8(.49)
Advanced	4	.80(.49)	6	.75(.25)	18	1.8(.49)	1	.50(.50)	10	1.3(.73)

[a] *M* = the mean number of comments per participant in the category.

SE = estimated standard error of the mean.

Expertise

Table 8.6 presents the relationship between situation awareness and expertise for novices and experts according to their level of expertise in both industry and in OCC.

The findings from Table 8.6 indicate a high concentration of responses at the elementary level of situation awareness for both novices and experts. This is consistent with the findings from Tables 8.4 and 8.5 and suggests that both novices and experts consider elementary information as fundamental for gaining situation awareness. Novices appeared to focus to a greater extent than experts on aspects classified at the core level. This may relate to an emphasis by novices on aspects such as gaps in the utilisation, weather, and crewing as a means of preparing for disruptions. However, experts may also have considered these aspects but not have verbalised all their observations. Further, there was little evidence to suggest that expertise in industry or OCC led to more advanced levels of situation awareness. Rather, it appeared that participants focused on elementary level aspects regardless of their level of expertise. The findings suggest that expertise may help participants gain a fundamental awareness of aspects such as passenger loadings or scheduling problems, but may not necessarily help them gain awareness of advanced aspects such as ways to limit consequences of disruptions or reduce operational constraints.

In conclusion, it appears from these findings that increased industry and OCC expertise of participants in domestic simulations may help them build better awareness of fundamental aspects of the domestic operations. However, there is little evidence to suggest that expertise helps participants achieve more advanced levels of situation awareness as the findings in the international simulations demonstrated. In this way, these findings differ from the international simulations.

Table 8.6 Frequencies and mean numbers of situation awareness comments by novices and experts according to the level of expertise in industry and in OCC. (*N*=33)

Situation Awareness	Novices in industry[a] and OCC		Experts in industry and novices in OCC		Experts in industry and OCC	
	n=8		*n=15*		*n=10*	
	f	*M(SE)*[b]	*f*	*M(SE)*	*f*	*M(SE)*
Elementary	8	1.0(.19)	15	1.0(.22)	8	.80(.20)
Core	19	2.4(.18)	41	2.7(.23)	27	2.7(.40)
Advanced	7	.88(.35)	21	1.4(.36)	11	1.5(.60)

[a] No participants were both novices in industry and experts in OCC.

[b] *M* = the mean number of comments per participant in the category.

SE = estimated standard error of the mean.

Decision Considerations

Participants took into account numerous and varied decision considerations prior to generating decision alternatives. The patterns that emerged from participants' comments led to the classification of these comments according to the degree of complexity into elementary, core, and advanced categories. For consistency, the method for classifying comments was the same as the method used in the international simulations.

Classification of Decision Considerations Comments

Comments were classified as elementary when participants considered fundamental aspects of the scenario such as aircraft commitments, passenger loadings, crewing commitments, and maintenance requirements. Comments indicating that participants accepted information without question were also classified as elementary. Comments classified at the core level indicated that participants identified further consequences of a situation or identified ways to manipulate resources to enable them to manage a situation. For example, they might have questioned whether a delayed flight was likely to have implications for crew duty hours or problems for connecting passengers, or they might have identified ways in which they could optimise the flight schedules in response to weather problems. Comments classified at the advanced level indicated that participants had considered ways to overcome the limitations of a situation. For example, they might have identified alternative ways to uplift passengers to avoid delaying a specific flight, or they might have negotiated to re-schedule maintenance work.

Aanalysis of Scenarios

The second stage of the domestic simulations required participants to consider the situations presented in three individual scenarios. The expert panel provided an optimum solution for each scenario in the domestic simulations and a set of decision considerations that participants should take into account during the decision-making process. This provided a means for measuring participants' responses in the scenarios. The scenarios were analysed according to participants' comments in relation to their revised situation awareness, information completeness, decision considerations, levels of expertise, decision-making styles, and generation of decision alternatives.

Scenario One

The following initial briefing was provided to participants at the start of Scenario One.

> The time is 1530. Pacific Southern airlines is operating Tokyo (NRT) to Sydney (SYD). However, the aircraft diverted to BNE for fuel and the ETA SYD is now 1715. This flight has 70 tranship passengers including 25 disabled people for Flight 826 SYD-ADL (Adelaide). The airport manager at Sydney says she definitely wants to hold Flight 826 for these tranships, as the number comprises half the booked load on Flight 826.

Scenario One required participants to consider the extent to which Flight 826 could be delayed to wait for passengers from a delayed inbound international flight. The aircraft operating Flight 826 was scheduled to operate back to Sydney to arrive before curfew, so participants also needed to take this aspect into account (see appendix). A further complication that needed to be identified and considered by participants was the arrival of the international flight at the International terminal in Sydney and the need to clear the 70 passengers through Customs and Immigration and then transport them all by bus to the domestic terminal which lay on the opposite side of the airfield. Participants were required to identify the considerations they took into account as part of their disruption recovery strategies. Figure 8.1 below provides a graphical depiction of this scenario and the flight display can be viewed in Appendix B.

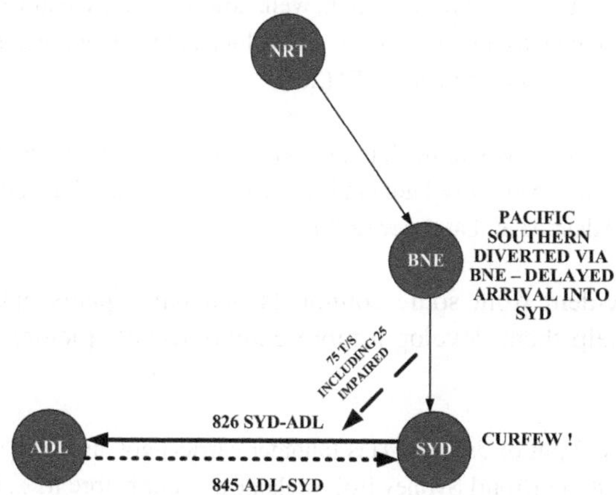

Figure 8.1 Graphical representation of domestic scenario one. See Appendix B for domestic utilisation (not to scale)

The optimum solution provided by the expert panel was to delay the flight as long as necessary, and manipulate several aircraft patterns to overcome a potential breach of curfew. The inter-rater reliability for classifying participants' comments into the categories of decision considerations in Scenario One was .76. The inter-rater reliability for identifying comments within each category was .80.

Revised Situation Awareness

In the familiarisation stage of the simulations, participants gained situation awareness through observation of information that was evident in the flight display and by requesting further information. Participants enhanced their level of situation awareness as information about the scenario was introduced and by requesting specific details in relation to the scenario problem.

Information Completeness

Nearly all participants (95 per cent) appeared to rely on the initial information provided for the scenario. This was in contrast to Scenario One in the international simulations, where many participants questioned the reliability and source of information. It was evident from the comments that participants appeared to make a quick assessment of the scenario and ask for information focusing on implications of the disruption. For example:

> Immediately, with a 1715 arrival I think, well, how long ... and that would be the question I would be asking of Sydney ... how long are we looking at to tranship them if indeed they do get in at 1715? (206)

> Then the first thing I would be doing is looking at the reservations system and [finding out] what loads have I got [and] what aeroplanes have I got going [from] Sydney [to] Adelaide at that time? (207)

It became evident from some comments that participants gathered specific information to help them develop a more comprehensive picture of a situation. For example:

> That's the last flight of the day? Has [another carrier] got any seats? ... Do we have another flight [from] Sydney [to] Adelaide an hour before it? ... I may also ring flight planning [to ask] what kind of high speed plans we can get ... and are the crew going to be in Sydney for an on-time departure? (208)

In Scenario One, there were 72 requests for further information by participants. These requests related mainly to passenger loadings (89 per cent of participants), airport handling (79 per cent), and crewing (58 per cent). In contrast, there were 29 requests in the familiarisation stage relating mainly to maintenance and weather aspects. These findings suggest that participants appeared to enhance their initial situation awareness by seeking information specific to the scenario. This supports the findings in the international simulations suggesting that situation awareness is a cumulative process.

Decision Considerations

Participants made 110 comments in relation to decision considerations in Scenario One. The comments were classified according to elementary (32 per cent), core (45 per cent), and advanced (23 per cent) categories.

Elementary Decision Considerations

Most participants (84 per cent) identified elementary decision considerations. Around half of the participants (58 per cent) accepted and acted upon the information given to them at the start of the scenario without challenging its content or source. A similar finding was evident in the international simulations. Although the scenario involved delaying a flight to wait for connecting passengers, participants focused on maintenance (47 per cent) and crewing (63 per cent) considerations. For example:

> ... I can see a swap that could be done between aircraft ... neither aircraft has got overnight maintenance required. We'd ring Maintenance Scheduling and check [their response to the situation] that if we did do the swap, we'd have NBV in Sydney and NBR in Perth. (209)

> What I'd be looking at ... is the crew going to Adelaide and back to Sydney? OK, so I'd be checking their hours ... if we have a delay of half an hour, would that be OK? (313)

The comments suggest that in domestic simulations, participants may regard specific aspects such as maintenance and crewing to be fundamental considerations regardless of the problem being considered.

Core Decision Considerations

Comments classified at the core level indicate that participants had an enhanced comprehension of the situation and considered ways to reduce the consequences of the situation, but only within the constraints of the scenario. For example, most participants (84 per cent) sought ways to mitigate the disrupted operation, such as shortening the ground times at airports, or asking if the flight could operate at a faster speed to recover some of the delayed schedule. In this way, participants could reduce the risk of arriving in Sydney after the curfew. For example:

> I would actually check the Sydney to Adelaide flying time and Adelaide to Sydney flying time [and] get a flight plan to see if there was a short time, [then] check the turn-around time in Adelaide, and see what they can do ... (314)

Participants were advised that a large group of transferring passengers were disabled. This situation added complexity to the scenario because of the increased uncertainty of the time involved transferring the passengers. It was evident that several participants (58 per cent) recognised the increased difficulty of transferring this group and the extensive connection times required. For example:

> Moving disabled people onto an aircraft takes ages. We would have to load them ... well we usually do load them before the other passengers ... 25 is a large number and [the transfer] is going to take ages. There is a very serious risk [that] if we did wait for them, we wouldn't get [Flight] 845 back into Sydney [due to the curfew]. (215)

It was evident from participants' comments that the decision considerations they took into account focused on resolving the situation within the constraints of the problem rather than considerations that may lead to more creative solutions. This suggests that decision considerations identified at the core level may limit decision outcomes.

Advanced Decision Considerations

The comments at the advanced level indicate that participants searched for ways in which they could avoid the situation, rather than work within the constraints. Most participants (68 per cent) identified one or more alternative solutions for the connecting passengers that would avoid the need to delay flight 826. For example:

> Do they have to go on [Flight] 826? Is there no other Sydney-Adelaide flight with seats available that we can put them on? Either that or can I send them via Melbourne and then across to Adelaide on another service? The next option is, can I send them directly Brisbane-Adelaide ... or can [I] send them again Brisbane-Melbourne-Adelaide? (202)

Other methods to avoid the situation included swapping aircraft patterns to prevent breaching the curfew in Sydney. For example:

> I'd be looking for another aeroplane in Adelaide that wouldn't be curfew affected. Now, if we go down the board, you have got NBW into Adelaide

> [at] 1945 [which then] goes Adelaide-Melbourne at 2020. I would be asking Crewing whether they could use that crew to come Adelaide-Sydney. (206)

It was evident that advanced considerations took into account ways to overcome restrictions having identified constraints that could be changed. For example:

> Well, it would affect our delay on [Flight] 826 which means we possibly couldn't do it. We could perhaps get a dispensation under extenuating circumstances from Sydney [authorities] to get back into Sydney. (101)

> So a scheduled number two engine change on that ... so then I have got to go cap in hand to Maintenance and say, well ... if we are in a curfew situation with NBR, how are you going to feel about me throwing out the ... engine change on NBW? (206)

There was substantial variation in the way participants considered the implications of the scenario as evident in the three categories into which decision considerations were classified. This was consistent with the approach participants used in the international simulations. To investigate whether this variation may be influenced by participants' levels of expertise, the study investigated the relationship between expertise and decision considerations.

Expertise

The analysis for the current section relies on the basis of ten years' experience to differentiate between novices and experts. This basis was established in the previous chapter. Although the information in the scenario was provided so that participants could decide whether or not to delay a flight, some comments from novices indicated that they regarded the delay as having been confirmed, rather than having to make the decision themselves. For example:

> Obviously the first thing is [that] you need to confirm times out of Brisbane and times into Sydney and obviously, probably more so with Sydney, with regards to the holding. Now obviously firstly they are going to hold [the flight for connection] and that's confirmed. (105)

Other novices appeared to identify several, separate aspects relating to a situation, often in rapid succession. These comments were interspersed with responses from the researcher:

> This one's out at 1845 and it's coming in when? Trying to have a look to see what's going where. That's the only Sydney [to] Adelaide flight ... what's our crew doing? And what are they doing the next day? So that's the old curfew back

> ... How many passengers are on [flight] 845? [are there] any other operators with a flight? I honestly don't know where I'm going with this. (104)

In the international simulations, novices appeared to focus on the surface features of a situation, using a 'bits and pieces' approach. Similarly, novices in the domestic simulation also focused on surface features. In contrast, experts appeared to form a more comprehensive picture of the situation by linking together various aspects such as numbers of connecting passengers, passenger loadings, the curfew limitation, and airport handling considerations. So it seems that experts are fairly quick to clarify information and able to integrate components of problems. For example:

> ... do we hold for the 70 passengers and the 25 wheelchairs? [With a] 1715 arrival [at the] international [terminal for an] 1805 [departure at the domestic terminal] there is no way they can do that in 50 minutes ... you've got a Sydney curfew problem at 2230 with the aeroplane coming back [from Adelaide] ... the first thing I would be doing is looking at the reservations system [to tell] me what loads have I got going Sydney to Adelaide at that time. (207)

Table 8.7 presents the relationship between expertise and decision considerations for novices and experts according to their level of expertise in both industry and OCC.

Table 8.7 Frequencies and mean numbers of decision considerations comments by novices and experts according to their level of expertise in industry and in OCC. (N=33)

Decision considerations	**Novices in industry[a] and OCC** *n=8*		**Experts in industry and novices in OCC** *n=15*		**Experts in industry and OCC** *n=10*	
	f	*M(SE)*[b]	*f*	*M(SE)*	*f*	*M(SE)*
Elementary	7	.88(.23)	13	.87(.17)	8	.80(.20)
Core	16	2.0(.63)	47	3.1(.22)	26	2.6(.37)
Advanced	5	.63(.26)	9	.60(.16)	10	1.0(.26)

[a] No participants were both novices in industry and experts in OCC.

[b] *M* = the mean number of comments per participant in the category.

SE = estimated standard error of the mean.

The data in Table 8.7 indicate that novices in industry and OCC appeared to focus on core considerations. It is also evident that participants with expertise at industry and OCC levels identified more core level decision considerations than novices. These findings suggest that participants focus on considerations of greater complexity than elementary considerations. Rather, they appear to focus on aspects enabling them to consider consequences of actions, and alternative ways to manage situations.

Decision-making Styles

The extent to which participants used rational and intuitive decision-making styles (DMS) during the decision-making process in the domestic simulations was examined. The analysis of data also examined the extent to which rational and intuitive decision-making styles differed between novices and experts.

Rational Decision-making Style

Similar to the classification of data in the international simulations, comments classified as low rational DMS indicated that participants showed little evidence of using a step by step approach. For example:

> At this stage for a 50 minute tranship [time] I'm not going to waste a whole heap of time ... worrying about it because at this stage it's OK. If [the international flight leaves Brisbane] later than 1530, that's when I'm concerned. Just my first thought would be ... well can we cancel something else and put on a supplementary flight? That would be one thing I would be looking at in the future. (208)

In contrast, comments classified as high rational DMS indicated that participants identified several considerations in a highly systematic way. For example:

> So there is not really the option of [getting the passengers] off in Brisbane and sending them some other way to Adelaide, so obviously what the next question would be if we run [flight] 826 that they are connecting with 25 minutes late and we take a standard turnaround in Adelaide, then we're only going to run 20 minutes late back into Sydney. We should be able to make curfew [at] 2250, so I would be asking Crewing what the crew are doing. (206)

Table 8.8 presents the relationship between participants' use of rational DMS and expertise according to the level of expertise in both industry and OCC.

Table 8.8 Classification of participants' use of rational decision-making style according to their level of expertise in industry and in OCC. (*N*=33)

	Novices in industry[a] and OCC		Experts in industry and novices in OCC		Experts in industry and OCC	
Rational DMS	*n*=8		*n*=15		*n*=10	
	n	%	*n*	%	*n*	%
Low	2	25	4	27	3	30
High	6	75	11	73	7	70

[a] No participants were both novices in industry and experts in OCC.

In Scenario One of the international simulations, participants appeared to use high rational DMS regardless of their level of expertise. According to Means, Salas, Crandall and Jacobs (1993), this approach could be expected in circumstances where time for decision-making is extensive. However, the data in Table 8.8 indicate that participants with expertise in industry and OCC used high rational DMS in the current scenario despite circumstances in which time for decision-making was limited. This is an important finding as it appears to contradict the findings from previous research arguing that rational decision-making cannot explain decision-making activities in situations where the goals are constantly changing (e.g., Klein 2001).

Intuitive Decision-making Style

The extent to which participants used an intuitive DMS was examined by classifying their comments as low intuitive DMS or high intuitive DMS. Comments classified as low intuitive DMS indicated that participants appeared to rely only to a limited extent on gut feeling. For example:

> Well, I would always leave [the decision] to as late as possible. I mean we want to get these passengers to Adelaide tonight. We don't want to put 70 people in the hotel. You can't take the [able] ones and leave the disabled ones behind. I mean that shouldn't really be an option. (312)

In contrast, comments classified as high intuitive DMS indicated that participants appeared to use a more complex, intuitive approach to the situation. For example:

> Going right back to my initial thought – my gut feeling – [with] 70 passengers, it's going to take a while to offload them, but I think 70 out of 150 is a significant amount of people. If it was ten and we could do that other thing, I would have

gone with that. But with 70, it's like half the flight so I think you've got to hold. (102)

You've got to figure out [that] if you're not going to get back into Sydney [due to the curfew], then you'll have bigger problems in the morning. I'd have a cut-off point, but I think with half an hour extra, they could probably do that [connection time]. (313)

Table 8.9 presents the relationship between participants' use of intuitive DMS and expertise according to the level of expertise in both industry and OCC.

The data in Table 8.9 indicate that participants with industry and OCC expertise seemed to use high intuitive DMS in Scenario One. The findings are consistent with research by Agor (1986) who contends that experts draw on intuition in times of uncertainty and when time for decision-making is limited. In contrast, novices appeared to use low intuitive DMS in Scenario One.

Table 8.9 Classification of participants' use of intuitive decision-making style according to their level of expertise in industry and in OCC. (*N*=33)

Intuitive DMS	Novices in Industry [a] and OCC *n*=8		Experts in Industry and Novices in OCC *n*=15		Experts in Industry and OCC *n*=10	
	n	%	*n*	%	*n*	%
Low	4	50	9	60	5	50
High	4	50	6	40	5	50

[a] No participants were both novices in industry and experts in OCC.

Generation of Decision Alternatives

The focus of the study was to examine the decision-making processes to the point where participants generate decision alternatives rather than evaluate the decision outcomes. In Scenario One, participants were required to determine whether to delay a flight to wait for passengers from another connecting flight. However, unless participants changed the flight patterns, an extensive delay would result in exceeding a curfew limitation. Most participants (84 per cent) agreed to delay the flight. In generating decision alternatives, it was evident that experts appeared to have better awareness of the curfew limitation than novices and were more likely to delay the flight provided that curfew was not exceeded. As a result, they appeared to formulate more workable strategies such as changing aircraft patterns or negotiating maintenance requirements to overcome the risk of breaching the curfew.

In contrast, novices commented that they would probably delay the flight, but it was not evident from their comments whether they were aware of the curfew, as their strategies did not appear to take this limitation into account. As a result, their strategies did not necessarily work, or at least required modifying to overcome limitations. The results support earlier findings in the study suggesting that experts appeared to know what might work and what the consequences of their actions were likely to be. Experts appeared more likely than novices to anticipate the occurrence of situations from which they could generate alternatives to provide workable solutions. In contrast, novices appeared to generate alternatives that addressed the immediate problem involving the connecting flight, but failed to consider the consequences of the delay adequately.

Scenario Two

The following initial briefing was provided to participants at the start of Scenario Two:

> The time is 1830. The aircraft NBO operating Flight 876 Melbourne (MEL) to Canberra (CBR) has radio'd back to Melbourne. He has indicated that he has an instrument indication problem. The captain says this problem has already been looked at in MEL and if same problem re-occurred, the aircraft would be grounded until such time as the problem was fixed. The captain wants to know whether he is to come back (to Melbourne) or keep going (to Canberra).

Subsequently, additional information was provided to participants:

> The time is now 1833. From the Melbourne engineers, the instrument component was changed in MEL prior to Flight 876. Clearly this has not worked and to fix it satisfactorily now would mean a four hour task. The aircraft will be AOG (aircraft operationally grounded) if it goes to CBR. Either way, it's out for the night. We'd prefer it back here to fix it properly. In addition, the engineer in Canberra hasn't done this sort of work before and there are no parts in CBR (to fix the problem).

Scenario Two was more complex than Scenario One and involved an aircraft that developed a mechanical problem shortly after take-off. Participants were required to consider the consequences of enabling the aircraft to continue to its destination or return to its departure port. Figure 8.2 provides a graphical depiction of this scenario.

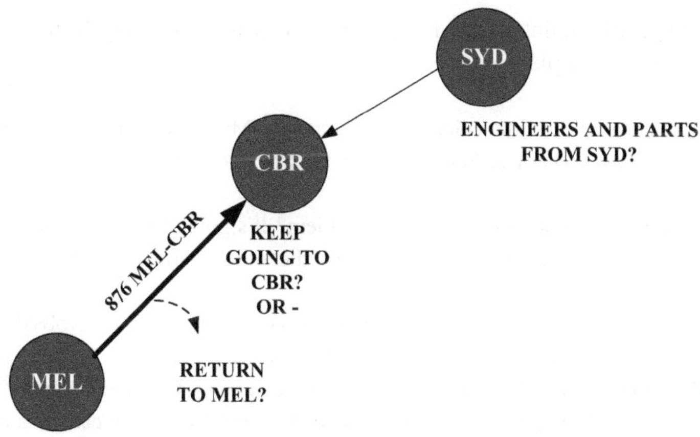

Figure 8.2 Graphical representation of domestic scenario two. See Appendix B for domestic utilisation (not to scale)

The optimum solution provided by the expert panel was to enable the aircraft to continue to its destination and manage the consequences of the disruption by identifying a range of alternative solutions. The increased complexity of this disruption required participants to take into account numerous decision considerations such as changing aircraft patterns, carrying passengers on alternative flights, delaying, adding, and cancelling flights and protecting flights limited by curfews or control tower staffing times. The inter-rater reliability for classifying participants' comments into the categories of decision considerations in Scenario Two was .65. The inter-rater reliability for identifying comments within each category was .73.

Revised Situation Awareness

Participants were made aware that each scenario in the domestic simulations was independent of the other scenarios. Accordingly, during Scenario Two participants enhanced their situation awareness as information about the scenario was introduced and by requesting specific details in relation to the unserviceable aircraft.

Information Completeness

An immediate dilemma identified by participants was the need to decide quickly whether to let the aircraft continue to its destination (Canberra), or return to its departure airport (Melbourne). Consequently, nearly all participants (95 per cent) requested specific maintenance information to assist in gaining further situation awareness. For example:

> Well I suppose I'd be immediately on the phone to Maintenance in Melbourne to find out exactly what was looked at in Melbourne ... (206)

> What was the synopsis on the initial problem? It's going to be grounded. Are there [maintenance] parts available? I'd ask this of the engineers. (103)

> Are we talking ten minutes? Are we talking hours? Is there a part involved that we need to now replace because it didn't work the first time ... ? Is that part available in Canberra? If not, how soon can we get a part up into Canberra? Is there another flight going to Canberra out of Sydney? Maybe we can get a part out of Sydney. We don't have to get it out of Melbourne. (202)

It appeared from participants' comments that they quickly comprehended the complexity of the situation. This was evident from additional information requested in relation to passenger loadings (89 per cent of participants) and crewing (79 per cent). For example:

> ... my initial question would be what capacity have we got Sydney [to] Canberra, to get the passengers back to Canberra? (206)

> What other resource do we have to try and get these people away? What other flights do we have available from Melbourne [to] Canberra for the remainder of the day with seats available? (301)

> Do any of these crews actually go in [to Canberra] and then turn around and come out the next morning? (304)

In Scenario Two, there were 116 requests for information by participants. Several observations are noteworthy from the data. The majority of participants requested information relating to maintenance aspects which was to be expected given the focus of the scenario. However, the emphasis on aspects such as passenger loadings and crewing supports the findings in Scenario One, suggesting that participants sought a body of relevant information that would enhance their awareness in relation to a specific situation.

Decision Considerations

Participants made 179 comments relating to decision considerations in Scenario Two. The comments were classified according to elementary (43 per cent), core (30 per cent), and advanced (27 per cent) categories.

Elementary Decision Considerations

Comments at the elementary level indicated that most participants took into account fundamental considerations such as aircraft, crew commitments, passenger loadings, and tranships. Most participants (89 per cent) focused on identifying available aircraft to replace the aircraft with the mechanical problem. For example:

> I guess my first thought was Ok, get this thing [flight 876] back. We are going to lose it for the day. What then is my next available aeroplane? I've gone straight to NBV which is available. However, it is doing a 1910 service to Adelaide for which I don't have another aeroplane. (210)

In addition, most participants (89 per cent) considered crewing related limitations. For example:

> I'm just thinking ... [Flight] 879 Melbourne to Hobart is going to be late whatever we do. Is the tour of duty for both crew[s] sufficient, so [that] we can delay the 879 Melbourne to Hobart [service]? (210)

As the disruption occurred to an aircraft operating on the Melbourne to Canberra route, it was not surprising that most participants (74 per cent) commented about the passenger loadings on that route. Some comments addressed alternative ways for passengers to travel. For example:

> I'm thinking that if we sent the aircraft to Canberra ... then our problem is in Canberra which would be [Flight] 879 ... [which has] 130 people on it. A quick glance down the [display] to see if we have got another [flight] going an hour later ... [Flight 889] is half full. That means we can get 70 [passengers on Flight 889, so] ... we've got 60 people left. Do we have any other Canberra to Melbourne [flights]? (208)

It was evident from the emphasis on aircraft patterns, crewing, passenger loadings, and tranships, that participants considered these aspects to be crucial in the decision-making process.

Core Decision Considerations

All participants took into account maintenance related considerations. The comments related specifically to the nature of the problem with the aircraft and to the level of available maintenance support. For example:

> We would have confidence in Maintenance watch and [would] get the aircraft [crew] to speak to maintenance watch. Now I'd be asking them ... [as] they know the defect: 'what's your prognosis?' Obviously if it's going to Canberra, and the

> [captain] reckons he's going to be [operationally grounded] there, have we got parts there ... with an engineer ... to get [the aircraft] serviceable? (301)

> ... [I'd also be] communicating with Melbourne [and asking] ... would it be an easy fix? Are we talking ten minutes, are we talking hours, is there a part involved that we need to now replace because it didn't work the first time. Is there a part available in Canberra? If not, how soon can we get a part up [there]? (202)

Another complexity of the disruption was the consideration for 40 connecting passengers to travel from Canberra to Adelaide via Melbourne. To address this problem, participants looked at alternative ways to transfer the passengers. For example:

> So I would be looking at ... how many seats have we got Canberra to Sydney at about the same time? [At] 1920 Canberra to Sydney, you have got 50 seats. Does that connect with anything Sydney to Adelaide? (206)

Comments at the core level indicated that participants took into account considerations beyond straightforward identification of problems such as aircraft commitments and passenger connections. For example, establishing the availability of spare parts and engineering staff indicates that participants were considering ways to resolve situations. The comments also emphasise the way in which participants worked within the constraints of the problem, such as moving passengers onto other scheduled flights to enable them to reach their destinations.

Advanced Decision Considerations

At the advanced level, participants sought ways to avoid rather than reduce the consequences of a situation. Many participants (68 per cent) indicated that they considered constraints such as curfews and control tower opening times but sought ways to avoid these constraints. This approach was not evident from comments at the core level. Participants also appeared to examine creative ways to manage the disruption in the scenario. For example, some participants considered interchanging types of aircraft resulting in complex negotiations to re-balance the aircraft types and arrange different crews. In contrast, responses at the elementary level related to less complex solutions such as changing aircraft patterns solely within one type of aircraft. In addition, participants looked for more innovative ways to provide aircraft to operate the flights. For example:

> I am also looking at the loadings to see if there's anything that has a light load that we could perhaps cancel, and [then we could] combine [flights]. (209)

> Could we create the circumstances [where] we can even swap with a wide body [aircraft] to free up an aircraft that will allow us to go Melbourne [then] Launceston [to] Hobart and probably back the same way? (204)

The information given to participants during the scenario advised them that the estimated time to resolve the mechanical problem was four hours. Several participants (32 per cent) challenged the accuracy of this information. They referred to previous experience in which maintenance estimations had fluctuated considerably. For example:

> The engineers told me four hours, but it could be half an hour. The guys could get in there and say, look ... it could be half an hour [to fix]. (105)

> What I am saying is ... if it's only an instrument indication problem, it could be a loose wire when they changed the instrument around. The guy gets in and [finds] the plug's come loose, just twigs it, and the aeroplane goes. [If that were the case] we [would have seen] it turn back [to Melbourne] with 150 people on it, for no reason whatsoever. (207)

> From what I've experienced, [the time of serviceability] can also go the other way, where a one hour fix can blow out to four hours. (101)

Despite the increased complexity of the scenario in comparison with Scenario One, the comments indicated the extent to which participants focused on fundamental aspects such as aircraft commitments, passenger loadings and tranships, and crewing. However, the comments at core and advanced levels demonstrated the emphasis that participants placed on specific aspects such as maintenance and ways in which they sought to overcome maintenance and regulatory constraints.

Expertise

In Scenario Two, some novices appeared to have difficulty assessing the situation and commented that they would seek help from others to handle the situation. For example:

> We've got a bit of time to play with, not much – sort of an hour or so. We can go through and have an assessment as to what we can do, have a look at what we've got and get whoever is on with me perhaps to do a brainstorm [to] see what we can come up with. [We could also] get suggestions from the guys actually on the front there and see what we can do. But at this stage that's my quick fix. (105)

Other novices appeared to assess the situation but had difficulty establishing what might work to address the problem. For example:

> I'm just looking for another flight out of Melbourne [to operate to] Sydney and hopefully connect this to Canberra from Sydney ... [but] I don't know if there is one there. So, I've got rid of 70 [passengers] haven't I? So, I've got to get rid of another 60. If I could get them to Sydney, I could possibly put them on [Flight] 849 if I could get them away. He's probably already gone hasn't he? That's right he has too. Possibly I'd be over-nighting the 60 others maybe. I don't know, not having much experience. (104)

In contrast, experts appeared to assess the situation quickly. They identified a number of consequences of the situation and began to devise contingency plans. It was evident that experts were able to comprehend the complexity of the situation better than novices. Their comments were often protracted, demonstrating their ability to identify numerous considerations that might influence the decision-making process. Further, experts could also explain what strategies might not work to resolve a situation. For example:

> Where's my Sydney [to] Adelaide ... doesn't help. I would have explored the possibility of sending the Adelaide [passengers] via Sydney actually on [Flight] 888 [as] there [are] probably enough seats on that. That doesn't appear to work. I think I would be looking to put ... [Flight 840] on the next available aeroplane which appears to be NBN, coming off Flight 839 from Adelaide at 2010. That would be an hour and a half delay on [Flight] 840. [The] crew have sufficient hours for that. (210)

As part of the decision-making process, experts appeared to identify readily constraints that might prevent a plan from working, as well as ways to reduce these constraints. For example:

> Now that aircraft is supposed to be in Sydney for an APU rectification. Given the gravity of the problem, I'd be negotiating with Maintenance for another slot on that APU. (210)

Given the increased complexity of the scenario, it was evident from the comments that novices experienced difficulty handling the maintenance problem. In contrast, experts were able to take into account the short time available for decision-making and were quick to respond to the problem. Table 8.10 presents the frequencies and mean numbers of decision considerations comments by novices and experts according to the level of expertise in industry and in OCC.

Table 8.10 Frequencies and mean numbers of decision considerations comments by novices and experts according to the level of expertise in industry and in OCC. (*N*=32)

Decision considerations	Novices in industry[a] and OCC *n=8*		Experts in industry and novices in OCC *n=14*		Experts in industry and OCC *n=10*	
	f	*M(SE)*[b]	*f*	*M(SE)*	*f*	*M(SE)*
Elementary	10	1.3(.25)	13	.93(.25)	11	1.1(.23)
Core	32	4.0(.27)	48	3.4(.36)	37	3.7(.40)
Advanced	11	1.4(.32)	21	1.5(.20)	18	1.8(.25)

[a] No participants were both novices in industry and experts in OCC.

[b] *M* = the mean number of comments per participant in the category.

SE = estimated standard error of the mean.

The data in Table 8.10 indicate patterns that differ from the findings in Scenario One. In Scenario One, participants appeared to focus on core level considerations. However, in Scenario Two it appears that all participants focused extensively on elementary considerations. This suggests that with the increased complexity of Scenario Two, experts may have needed to establish fundamental aspects such as selecting available aircraft commitments, passenger tranships, and crewing.

Decision-making Styles

Rational Decision-making Style

In Scenario Two, a few participants (11 per cent) appeared to use low rational DMS. Their comments suggested little evidence of using a step by step approach. For example:

> 1850, so we've got heaps of passengers to move. It is huge. Right, we can offload a few to [Flight 889] here. (104)

> ... OK, maybe ... if we decided that we want to cancel that flight, I'd be looking at getting some capacity between Melbourne and Sydney because we've got a Flight 849 that's going Sydney [to] Canberra. (304)

In contrast, 89 per cent of participants appeared to use high rational DMS. The comments demonstrated that these participants took into account the decision considerations in a highly systematic way as they assessed available aircraft, maintenance, crewing, and regulatory aspects. For example:

> Alright then, we have a major disruption ... we have got the engineers organised and Melbourne airport advised regarding slot availability ... There is no emergency. [This has been] established from the crew. The approach and landing will be normal ... we then have to look after the 150 people that are on the ground with [Flight] 876 and 130 coming out of Canberra [for] Melbourne. (101)

Table 8.11 presents the relationship between participants' use of rational DMS and expertise according to the level of expertise in both industry and OCC.

The results in Table 8.11 support the findings in Scenario One despite the greater complexity of the disruption in Scenario Two. Regardless of their level of expertise, it is evident that participants use high rational DMS during decision-making in complex situations. A similar result in the international simulations could be accounted for by the broader time-frame for decision-making in international operations. However, the time for decision-making in domestic operations is considerably reduced. Therefore, it might be expected that participants would not have had time to use high rational DMS. It has generally been found that decision-makers have insufficient time in complex environments to compare options. However, the results in this study are in contrast to these findings. It appears that in some complex, dynamic decision-making environments such as OCCs, participants may well use rational decision-making styles.

Table 8.11 Classification of participants' use of rational decision-making style according to their level of expertise in industry and in OCC. (*N*=32)

	Novices in industry[a] and OCC		**Experts in industry and novices in OCC**		**Experts in industry and OCC**	
Rational DMS	***n*=8**		***n*=14**		***n*=10**	
	n	%	*n*	%	*n*	%
Low	1	13	-	-	-	-
High	7	87	14	100	10	100

[a] No participants were both novices in industry and experts in OCC.

Intuitive Decision-making Style

A few participants (21 per cent) appeared to use low intuitive DMS in Scenario Two as there was little evidence to suggest that they used their intuition to any extent. For example:

> ... I'm now looking to see how to move the people out of Canberra and out of Melbourne. I would return that aircraft and I would ask them to rectify the aircraft as soon as possible and I would continue with the flight ... (314)

In contrast, most participants (79 per cent) appeared to use high intuitive DMS. It was evident that these participants considered many alternatives concurrently and relied on their gut feeling to determine which decision considerations might work better than others. The comments demonstrated how participants might also take into account the considerations of other airline departments. For example:

> Canberra aren't going to like this very much because I'm splitting the load. They would probably want me to run [Flight] 879 off the back of 889 and take the delay on 889 till later which is 2030. I think I will go with my original way at this stage [as] I have got a couple of options open to me depending on what the port want to do ... (208)

The comments also demonstrated the way in which participants might use intuition to evaluate decision alternatives. For example:

> Is there any way we can get him to Sydney? OK, we've only got one flight there but that's carrying many [passengers]. Can it be fixed in Sydney? [Are there] parts [in Sydney]? [I'm just trying] to see whether it's going to be worth the risk or not. Sometimes it's gut feeling. Sometimes it [depends] on who the captain is. (304)

Table 8.12 presents the relationship between participants' use of intuitive DMS and expertise according to the level of expertise in both industry and OCC.

Previous research has examined the use of intuition in complex situations with the result that intuition appears most successful in conditions where a high level of uncertainty and ambiguity exists and where time is limited to process a wealth of information (e.g., Sadler-Smith and Shefy 2004a). The focus has been extensively on experts and their use of intuition. However, the data in Table 8.12 suggest that the use of intuition for *both* novices and experts in OCCs may be very important to decision-making, in particular to help resolve complex disruptions.

Table 8.12 Classification of participants' use of intuitive decision-making style according to their level of expertise in industry and in OCC. (*N*=32)

	Novices in industry[a] and OCC		Experts in industry and novices in OCC		Experts in industry and OCC	
Intuitive DMS	*n=8*		*n=14*		*n=10*	
	n	%	*n*	%	*n*	%
Low	3	38	2	14	-	-
High	5	62	12	86	10	100

[a] No participants were both novices in industry and experts in OCC.

Generation of Decision Alternatives

As participants generated decision alternatives, it became evident that experts devised more sophisticated strategies than novices to recover from the disruption. For example, rather than delaying several flights they indicated that part of their strategy was to continue negotiating the use of the unserviceable aircraft, despite advice that this aircraft was not likely to be available again for the remainder of the night. These strategies enabled experts to consider a wide selection of decision alternatives.

In contrast, novices appeared to seek straightforward solutions. For example, they commented that they would reschedule the disrupted flights using the first available aircraft, or that they would carry passengers whose flights were disrupted on the next available flights. This suggests that novices considered decision alternatives that enabled them to address the situation but limited the scope of the solutions.

Scenario Three

The following initial briefing was provided to participants at the start of Scenario Three:

> The time is 0600. There is an alternate[1] on Melbourne until 0930. There is possible fog, but as yet no sign of any fog. There is also an alternate on Canberra until 1130 due fog. Some fog patches have been sighted.

1 Meaning that aircraft operating to MEL must carry sufficient fuel to be able to divert to a suitable airport if they could not land in Melbourne due to the weather.

Additional information was provided soon after:

> The time is now 0645. The captain of the training aircraft NBN says that the weather at Avalon (AVL) (near Melbourne) was not good. The airport is nearly closed with thick fog. The captain wants to wait for a while to see if the weather at Avalon improves. He says Melbourne is clear at present but fog is a strong possibility.

The following request was made immediately after the previous advice:

> Pacific Southern which is operating internationally into Sydney (SYD) has been delayed extensively. They now wish to terminate their flight at SYD rather than operate the aircraft SYD-MEL-SYD. They are due into SYD at 1405. They are requesting a charter aircraft from us to operate their SYD-MEL-SYD service between the international terminals at both ports. They want the same aircraft to operate both flights and they want to operate to the following schedule: Depart SYD for MEL at 1530 with 245 passengers and depart MEL for SYD at 1330 with 205 passengers. They need a response from us by 0730.

Scenario Three was the most complex of the domestic scenarios and simulated two concurrent events: weather situations (fog) at two airports and a request to schedule an aircraft to operate two charter flights. Figure 8.3 provides a graphical depiction of this scenario.

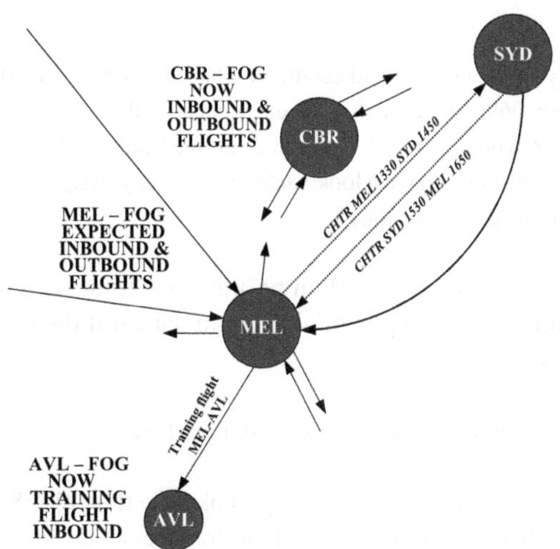

Figure 8.3 Graphical representation of domestic scenario three. See Appendix B for domestic utilisation (not to scale)

Similar to the third scenario in the international simulations, the expert panel did not provide an optimum solution for this scenario due to the high complexity of the scenario and the numerous actions that participants could take. Instead, the panel identified several considerations that participants should take into account during the decision-making process such as deferring the charter request until time was available to consider it fully, seeking high fuel uploads on specific flights, and delaying departures to airports with poor weather conditions. The inter-rater reliability for classifying participants' comments into the categories of decision considerations in Scenario Three was .71. The inter-rater reliability for identifying comments within each category was .83.

Revised Situation Awareness

In addition to requests concerning specific details about the fog situation in Canberra and Melbourne to enhance their situation awareness, participants also requested information in relation to the proposed charter flights.

Information Completeness

As the scenario primarily involved a weather situation, many requests for further information related to weather conditions. Some requests for information appeared to be based on participants' previous experience with handling weather situations. For example:

> ... I'd probably ring Canberra and ask them what their opinion of [the weather] is and if it's only 0600, it's probably too early because the sun is not up. My next question of them would be to ask the crew ... on the Canberra [to] Sydney [flight at] 0645 ... [if they could] have a look backwards [after getting airborne] and tell us ... the extent of [the fog]. (206)

Other comments related to ascertaining the need to operate a training flight. Participants identified this as a potential flight to cancel if the weather disruption worsened. For example:

> ... and the training must be done? It can be deferred? (202)

> I'd just check with the captain ... so I'd say 'look mate [Flight] 858 is obviously [scheduled to depart] after you [return from training so] do you think you're going to be able to get the training completed if you can do it out of Melbourne?' (105)

Several participants appeared to have some difficulty understanding the requirements of the charter flights and asked for clarification. For example:

> Just looking initially there's no aircraft to operate that. We've already got a Sydney [to] Melbourne flight operating at the same time. Was that Sydney [to] Melbourne or Melbourne [to] Sydney? What time is it despatching ... is it arriving back in to Melbourne at 1530? (103)

> So does our charter have to depart [from] Melbourne ... we can't come in from somewhere else? (104)

In Scenario Three, there were 202 requests for information by participants. Participants requested considerable information primarily about crewing (95 per cent of participants), passenger loadings (95 per cent), and weather (89 per cent).

Decision Considerations

Participants made 206 comments relating to decision considerations in Scenario Three. The comments were classified according to elementary (30 per cent), core (41 per cent), and advanced (30 per cent) categories.

Elementary Decision Considerations

Comments classified at the elementary level indicated that participants appeared to identify basic level considerations in relation to the weather situation such as aircraft patterns, tranships, and crewing. Many of the crew considerations related to the consequences of cancelling a training flight. For example:

> [I'd look at] either of two things. Either [I could look at] holding training and cutting it short for that period of time. They (Crewing) might be able to get half the candidates done running around Melbourne and Avalon. Training could still be done at Avalon if Avalon clears. [Or, I'd look at] sending it to another port and hoping Melbourne clears around the 1015 [a.m.] mark. (103)

> I'd talk to the Crewing people [to suggest] that the training [flight] could be [cancelled] or may not be completed for every one [of the trainee pilots]. I'd be asking the Crewing people what effect this [may] have on the training. (312)

The request for the charter flights required participants to provide a particular type of aircraft to operate the flights. Comments at the elementary level suggested that participants appeared to look initially for an easily identifiable option, such as an available wide-body aircraft. If none was found, some participants (32 per cent) considered alternative ways to carry the passengers such as placing them on existing flights, thereby negating the need for the charter. Other participants (11 per cent) stopped searching for options due to the priority given to handling the fog situation. For example:

> It's not looking too good anyway at that time. I am looking at these famous gaps – what I can do to accommodate [the passengers] ... I don't think I can get them a charter if this is the limit of our fleet. We have got nothing in maintenance I can see on the [display] anywhere. The other way I am looking at it, is whether we could shift these 90 off and distribute them across other [flights]. (204)

> It's very hard to commit [to a schedule for the charter], especially with this unknown fog in Melbourne, but if they want to pursue it a little bit more later on, just give me another couple of hours. (202)

While many participants (58 per cent) took into account basic level considerations such as crews and tranships, others (16 per cent) indicated that they would not consider alternative actions until flights became disrupted due to the weather situation.

Core Decision Considerations

The weather problems in the scenario were at Melbourne and Canberra. Therefore, it was not surprising that most participants (79 per cent) readily identified any flights that were to operate to Melbourne and Canberra. They also sought further information in relation to alternate airports that could be used if Melbourne or Canberra airports became unavailable. Identifying specific considerations enabled these participants to prepare themselves better for likely disruptions. Participants also considered precautionary measures such as loading additional fuel on flights operating to Melbourne and Canberra. For example:

> I'd have a look at ... what aircraft are heading into Melbourne, what aircraft are heading into Canberra [and] what possible implications that has on the rest of the day if there's fog. (104)

> ... I would probably ring Flight Planning and make sure [the aircraft] are ... all gassed up to max[imum] tanks at this stage. I wouldn't offload people at this early stage ... I'd just say, carry extra fuel or something like that. (208)

In relation to the request for the charter flights, comments at the elementary level indicated that participants looked for straightforward ways to process the request. In contrast, responses at the core level indicated that participants tried to negotiate terms of the charter. For example:

> ... seeing even if I can make a hole in our schedule, that we can approach them with a different time. Well I think that would be the way we are thinking ... what time do you want it? They tell us what time they want it and we tell them we can do it at this time. How does that suit you, or not at all? (206)

Many participants stated that they would not disrupt airline booked passengers in order to satisfy a request for the charter. For example:

> [I'm] just looking up what our aircraft are doing [at] that particular time. We probably wouldn't look at cancelling one of our services so that we can operate a charter. We don't have an aircraft available to do anything out of Melbourne at 1330 without affecting our own services. (209)

> One thing with charters for other people – we always put our schedule first, so if it looks like it's going to ... cause a major disruption to our existing schedules, that's always a pretty black mark against being able to do it. (215)

It was evident from participants' comments that the considerations taken into account focused on resolving the situation within the constraints of the problem. This may help to explain the reluctance of participants to consider more creative solutions without disrupting passengers already booked on flights. This focus was evident in previous scenarios and suggests that participants who only take into account elementary or core level considerations may not identify solutions that overcome problem constraints. As a result, decision-making outcomes are likely to be sub-optimal.

Advanced Decision Considerations

Comments at the advanced level indicate that participants looked for innovative ways to minimise the effects of the weather situation. For example, several participants (53 per cent) indicated they would prevent the training flight from departing or they would cancel the flight specifically to make the aircraft available for use in the weather disruption. This is in contrast to comments at the elementary decision considerations level which related to an inability to operate the flight due to the weather conditions. Participants considered stopping flights from departing, combining flights to make available a spare aircraft, and protecting gaps that could be used to cover further disruptions. For example:

> I saw straight away that the obvious choice was that [Flight 830] going Melbourne to Adelaide. There's also a wide body [aircraft] WBA going Melbourne to Adelaide ... [A total of] 45 and 175 means that that aircraft can be full. This one [Flight 830] doesn't have to go. [This] creates a spare aircraft for a short period. (102)

> [I'll] see how this fog starts developing. I am going to get into all sorts of trouble and I'll need all the gaps I can [have] in the schedule in order for me to keep my own schedules. (202)

Many participants (42 per cent) were concerned to verify the reliability of information concerning weather conditions. For example:

> [This] makes me wonder where the information on the weather [is] coming from. Is it coming from the Flight Planning section or is it coming from the people on the ground in Canberra? (206)

> No, if he's saying that Avalon [airport]'s bad, there is absolutely no point going there. I'd be saying 'what's your view of the forecast? Is it going to stay bad? Is the training in fact going to be a lost cause'? (210)

In response to the request for the charter flights most participants (89 per cent) considered manipulating aircraft types. This approach involved greater complexity than a possible solution using one aircraft type only. Other participants recognised further operational difficulties as a result of organising the charter. For example:

> It's an international service and they won't [handle] Melbourne domestic [terminal] to Sydney ITB (international terminal building). Normally we don't agree with that [operation] because the ports can't handle the situation with [a flight] going from the other [terminal]. You've got customs problems [and] you've got immigration problems, because you are going from a non-sterilised area to a sterilised area, and you'll have cross feeding of passengers. (207)

Comments at the advanced level indicate that some participants appeared to anticipate events and addressed the increased complexity of the scenario by considering the consequences in great depth. In doing so, they looked for opportunities to generate creative solutions by identifying gaps in the utilisation that could assist them to solve disruptions, while protecting specific flights and checking to ensure that information upon which they relied was credible. The initiatives demonstrate a high level of creativity which was not evident from comments at the core level.

Expertise

In Scenario Three, it appeared that some novices experienced difficulties coping with the complexity as the circumstances in the scenario changed rapidly. For example:

> ... once he's on the ground then I've got ... other issues. [It's] just [that] if he's on the ground then I know exactly where I'm standing with what flights ... (104)

Other comments demonstrated ways that some novices handled these complexities. For example:

> Prioritise ... that's kind of what I do. The flights that are up in the air are the priority. The ones on the ground are the pending questions. [Questions] about charters get dropped to the background. (101)

In contrast, experts appeared to comprehend the events in the scenario better than novices. They used the information provided and assessed the consequences of the situation quickly. Further, they indicated a series of actions they would take immediately. For example:

> So that means any aircraft operating to Melbourne will require an alternate destination. Alright, I would have a look at all flights operating to Melbourne and I would ask what holding fuel they have. (314)

> Generally in this situation, it's the arrivals that are going to be the problem. So I'd be looking at all arrivals up to 0930 and asking the question: where do we want the aircraft to go if it can't get into Melbourne? (312)

In contrast to novices who were quick to reject the request for charter flights, experts appeared to consider the request more extensively by investigating the availability of crews and the feasibility of changing aircraft patterns. While no participants satisfied the request, some experts continued to look for opportunities to provide the charter as the scenario progressed. For example:

> Can't always do [a charter] ... what we could offer them off the top of my head and I'm still thinking about it, is a back-to-back charter Sydney – Melbourne – Sydney ... I'm now looking at the possibility of giving them a wide-body jet and ... trying to see if I can ... shunt the narrow body on the wide-body schedule, thus enabling us to do a charter out of Sydney. (314)

These comments provide further evidence of the differences in the way that novices and experts manage a situation. Table 8.13 presents the frequencies and mean numbers of decision considerations comments by novices and experts according to the level of expertise in industry and in OCC.

The data in table 8.13 indicate that participants appeared to focus on core level considerations regardless of their level of expertise. In contrast, participants in the previous scenario appeared to focus on elementary considerations. This suggests that in highly complex situations, participants may regard core considerations as vital for decision-making, rather than the basic considerations evident at the elementary level. The result also contrasts with the findings from Scenario Three in the international simulations where participants appeared to focus on advanced level considerations. An explanation for this variation may be that participants in the domestic simulations may not have had sufficient time to take into account advanced levels of considerations due to the time constraints.

Table 8.13 Frequencies and mean numbers of decision considerations comments by novices and experts according to the level of expertise in industry and in OCC. (*N*=33)

Decision considerations	Novices in industry[a] and OCC *n*=8		Experts in industry and novices in OCC *n*=15		Experts in industry and OCC *n*=10	
	f	*M(SE)*[b]	*f*	*M(SE)*	*f*	*M(SE)*
Elementary	10	1.3(31)	20	1.3(.29)	13	1.3(.21)
Core	16	2.0(.50)	36	2.4(.29)	33	3.3(.50)
Advanced	29	3.6(.53)	43	2.9(.43)	36	3.6(.52)

[a] No participants were both novices in industry and experts in OCC.

[b] *M* = the mean number of comments per participant in the category.

SE = estimated standard error of the mean.

Decision-making Styles

Rational Decision-making Style

In Scenario Three, 95 per cent of participants made comments that were classified as high rational DMS. The comments indicated that these participants were highly systematic in the way in which they considered the two concurrent events. For example:

> Ok, first of all I look at what aeroplanes are going ... in and out of Canberra and Melbourne in the early hours of the morning up until 0930 ... you've got the training flight – that's not a real problem. You've got [Flight] 821 out of Canberra from Melbourne at 0800 arrival. You've got Launceston inbound [Flight] 870 at 0800. You've got [Flight] 831 from Adelaide. (207)

> Right, 1330 out of Melbourne. What I am doing now is just looking at what availability [of] an aeroplane sitting on the ground in Melbourne that can do [the charter]. For a start, I don't disagree that we won't [operate] Melbourne

> domestic [terminal] to Sydney [international terminal building] ... you've got customs problems, you've got immigration problems ... (207)

Table 8.14 presents the relationship between participants' use of rational DMS and expertise according to the level of expertise in both industry and OCC.

The data in Table 8.14 indicate that participants appeared to use high rational DMS despite the limited time for decision-making. The findings are consistent with the findings for the use of rational decision-making style in the previous two scenarios. The conclusion drawn from these findings is that both novices and experts appear to use a high rational decision-making style in complex, dynamic environments such as OCCs despite previous literature suggesting the unsuitability of rational decision-making in these environments.

Table 8.14 Classification of participants' use of rational decision-making style according to their level of expertise in industry and in OCC. (*N*=33)

	Novices in industry[a] and OCC		Experts in industry and novices in OCC		Experts in industry and OCC	
Rational DMS	*n=8*		*n=15*		*n=10*	
	n	%	*n*	%	*n*	%
Low	2	25	2	13	1	10
High	6	75	13	87	9	90

[a] No participants were both novices in industry and experts in OCC.

Intuitive Decision-making Style

Nearly all participants (95 per cent) made comments that were classified as high intuitive DMS. It was evident from the comments that these participants seemed to know what might work best to manage the complex weather disruption. For example:

> Alright, so we have got him possibly diverting to Hobart. That is a little bit better. We got rid of our passengers on [Flight] 804. We either put them off until [Flight] 808 at 1200, or if we get back in [to Melbourne] ... we could swap [that aircraft] for a 1030 departure, bring [the passengers who diverted to Hobart] up at 1030, and that should still get us back in there in time to do [Flight] 811 at 1600. (215)

> There's no rule. It's something that you think in your brain – do I hold [this aircraft] or not? This one's in the air, this one's just about to [depart, so] do I

> hold him here until the weather improves? I might say there's no point in [that aircraft] getting airborne ... I'd be looking at holding [all the aircraft] for at least half an hour to see how it goes, because you've got this fog rolling in, in Melbourne ... (207)

Table 8.15 presents the relationship between participants' use of Intuitive DMS and expertise according to the level of expertise in both industry and OCC.

The data in Table 8.15 support the findings regarding the use of intuitive decision-making style in Scenario Two in the domestic simulations. The conclusion drawn from the findings is that both novices and experts are likely to use a high intuitive decision-making style to manage disruptions in highly complex situations.

Table 8.15 Classification of participants' use of intuitive decision-making style according to their level of expertise in industry and in OCC. (*N*=33)

	Novices in industry[a] and OCC		Experts in industry and novices in OCC		Experts in industry and OCC	
Intuitive DMS	*n=8*		*n=15*		*n=10*	
	n	%	*n*	%	*n*	%
Low	2	25	2	13	1	10
High	6	75	13	87	9	90

[a] No participants were both novices in industry and experts in OCC.

Generation of Decision Alternatives

It was evident that all participants encountered considerable difficulty managing the complexity of both the weather and the charter flight situations in this scenario. In terms of the weather situations, experts were proactive, devising strategies to minimise or avoid disruptions such as requesting additional fuel on flights to enable them to circle for considerable time at a fog-bound airport without having to divert to another airport. In contrast, novices appeared to identify specific problems such as aircraft patterns likely to be disrupted but were more likely than experts to wait for a disruption to occur before considering further action. Novices appeared ready to base quick decisions such as cancelling the training flight on limited information.

Although no participants could satisfy the requirement to provide the charter, experts continued to investigate ways to achieve this despite the complexity of handling simultaneous problems. In contrast, novices readily dismissed the charter and focused on the weather situations. The findings were consistent with the previous scenarios, highlighting the more advanced decision-making processes of experts to manage disruptions in complex circumstances.

Summary of Trends Across the Domestic Scenarios

A number of patterns emerged from examining decision-making in the three scenarios. In complex situations, participants sought information specific to the problem in the scenario similar to the response of participants in the international simulations. However, it was also evident that participants in the domestic simulations sought a broader body of information enabling them to enhance their level of situation awareness. Second, as complexity increased in terms of the three scenarios in domestic simulations, it was evident that participants appeared to consider more core and advanced decision considerations. This suggests that creating ways to reduce or avoid the consequences of situations rather than managing within constraints may be more important for participants where complexity is high. This pattern was evident for both novices and experts and suggests that expertise of participants may not be sufficient to explain why some controllers in OCCs focus on elementary aspects of a disruption, and others focus on advanced aspects. Third, in complex situations it was evident that novices experienced difficulty assessing the situation and coping with the problems, especially as circumstances changed during the scenarios. In the most complex scenario, both novices and experts focused on core level decision considerations. This was inconsistent with the findings in the international simulations and suggests that time constraints for decision-making in the domestic environment may limit OCC controllers' ability to recognise advanced considerations such as creative ways to solve operational problems. Fourth, it was evident in the domestic simulations that novices and experts used a high rational decision-making style despite the complexity of the scenarios. This result was surprising and contrary to previous arguments that the rational decision-making style has limited applicability in complex situations and changing circumstances.

Summary

This chapter presented the findings and analysis in relation to decision-making processes of participants during the 19 domestic simulations. The chapter was structured to replicate the structure of the previous chapter enabling ready comparisons of data between the international and domestic simulations.

Participants in the domestic simulations sought a higher level of preparedness for disruptions than participants in the international simulations. This was demonstrated by their focus on identifying sources of potential disruption and areas of weakness in the schedules that could be exploited to assist problem solving. In relation to acquiring situation awareness, the focus by participants predominantly on elementary and core level information suggests that they considered this information as fundamental to decision-making in domestic OCCs. It also appears that participants' industry experience was more beneficial than OCC experience in terms of gaining situation awareness in domestic OCCs.

Chapter 9
Conclusions and Implications for OCCs

Introduction

This chapter draws the main themes and results together and gives some direction for further work in this area. The chapter includes a summary of the findings, conclusions reached, and a discussion of the contributions of the study to theory and practice. Finally, the chapter presents limitations of the study that need to be taken into consideration which identify areas for future research.

Summary of the Study

The study conceptualised airline operations controllers as decision-makers who are responsible for monitoring flight operations, identifying potential and actual disruptions, and generating alternative courses of action. As described earlier, the differences in decision-making environments for international and domestic OCCs are quite pronounced. International environments typically involve a low volume of flights, long flight stages which may take place during the day or night, and extensive ground times between flights. As a result, decision-making may take place over extended time periods.

In contrast, domestic environments are typically more intricate as they are characterised by a high volume of flights, short flight stages, minimal gaps between flights, nightly maintenance requirements, and complex crewing commitments. As a result of these differences, decision-making in domestic environments is generally far more intense than in international environments, requiring controllers to make numerous decisions often within limited timeframes. The study conceptualised differences in decision-making according to these environments.

Controllers in airline OCCs rarely have time during airline disruptions to explain their reasons for decision-making. This means that there is considerable difficulty examining decision-making processes for disruptions that occur in real life; yet capturing controllers' thought processes was central to the study. This difficulty was addressed by designing simulations that replicated real-life operational disruptions. The simulations provided a means for investigating and comparing decision-making processes of participants in both international and domestic OCC environments. Thus, the method mimicked a naturalistic decision-making environment as closely as possible.

Both qualitative and quantitative data were collected. The use of qualitative methods enabled the collection of rich, in-depth data which provided opportunities to gain individual points of view. This approach was central to the study as it enabled the researcher to identify ways in which participants gained situation awareness and recognised decision considerations for resolving disruptions. Personal and professional demographic data were collected to enable comparisons to be made of frequencies between different groups and the frequencies of responses classified in various categories were calculated. However, the dominant approach for the study involved analysis of qualitative data.

Think-aloud protocol was used to gain an insight into participants' thought processes as they managed the disruptions during the simulations. This method was particularly appropriate for the study because controllers in OCCs do not have time during actual disruptions to explain their reasons for querying or requesting information, or taking certain actions. So using this method encouraged participants in the simulations to express their thoughts. In this way, the study provided access to underlying thought processes, reasoning, and behaviours likely to be used by OCC controllers in real-life. The study was conducted by adopting a multiple case-study approach, which enabled more information to be gathered, provided a deeper understanding of the decision-making process in OCCs, and increased the generalisability of the research.

Conclusions

Situation Awareness and Information Completeness

The relationship between situation awareness and information completeness was examined both during the familiarisation stage and throughout the simulations. In the familiarisation stage, participants' comments indicated that they acquired a level of situation awareness by observing and identifying several aspects of the flight displays. However, it appeared that the level of situation awareness gained may have been inadequate to prepare controllers for handling disruptions. As a result, participants requested further information. The analyses of comments provided insight into ways in which participants perceived particular elements of the flight displays to gain situation awareness and it was evident from their questions and comments that some elements were more prominent than others. The classification of elements according to the degree of complexity does not appear to have been made in previous studies. Nor is there explicit mention of differences between the perceived elements within level one of Endsley's (1995a) model as discussed earlier in Chapter Four. In the study, the recognition of differences between the perceived elements led to the classification of participants' comments according to three categories of situation awareness: elementary, core, and advanced.

Participants in both the international and domestic simulations focused quickly on key relevant information from the flight display. In the international simulations, this information related predominantly to core and advanced level sub-categories of situation awareness. This meant that these participants focused on information likely to help them identify potential problems and likely solutions. For these participants, information at the elementary level appeared only to provide superficial situation awareness. In contrast, participants in the domestic simulations focused extensively on elementary level sub-categories to provide just a fundamental level of situation awareness. Participants in the domestic simulations also sought more information within each sub-category of situation awareness than participants in the international simulations. As a result, it appears that the accumulation of information may be more critical for gaining initial situation awareness in domestic OCCs than in international OCCs.

During the simulations, participants' questions and comments indicated that they enhanced their initial situation awareness by gathering content-specific information. In the international simulations, participants emphasised the need to accumulate current and accurate information that was reliably sourced. Accordingly, they often re-checked the information provided. During the domestic simulations, participants' comments and actions indicated that time for decision-making was very limited. This was evident from the reliance they placed on the information provided as a basis for quick assessment of the situation, rather than challenging the source or reliability of information. In complex disruptions, the participants appeared to have difficulty coping with the extensive amount of information, resulting in some loss of situation awareness during the disruptions. Therefore, in complex situations it seems that OCC controllers are likely to become overwhelmed with the volume of information received and may not have time to question its source or reliability. This is likely to be detrimental to their decision-making processes.

A number of conclusions can be drawn from these results. First, the results suggest that situation awareness is a cumulative process. Second, it appears that situation awareness may be acquired in distinct stages. In the study, this occurred during the familiarisation stage where the environment was static, and then during the simulations as the environment changed. Third, participants appeared to gather information until they reach a certain level of initial situation awareness. Therefore, faced with complex situations, a high volume of information appears to detract from situation awareness as controllers have limited time to process it. This loss of awareness appeared to occur to a greater extent in the domestic rather than in the international simulations. As a result, acquiring sufficient information as a basis for gaining situation awareness may be more important in the familiarisation stage rather than during disruptions. Fourth, acquiring information in the familiarisation stage may be more important for controllers in domestic rather than international OCCs due to the limited time for decision-making in domestic OCCs. This suggests that controllers in domestic OCCs may need to prepare for disruption handling more comprehensively than controllers in international OCCs and highlights the importance of clear, comprehensive, and timely briefings.

Situation Awareness, Experience, and Expertise

The study examined the relationship between situation awareness and experience at two levels: the number of years in which participants had been in the industry and the number of years in which they had been in the OCC domain. In the familiarisation stage of both the international and domestic simulations, participants' comments indicated that they emphasised aspects relating to their own previous industry backgrounds. This alludes to the importance of past experience in terms of influencing the way in which participants may brief themselves prior to a disruption. While experience may enable participants to focus their attention on familiar information, limited attention to broader aspects may result in participants' failure to gain overall situation awareness. This raises an issue for OCC management to be aware of potential biases as a result of controllers' previous experience and for the need to develop training methods that assist controllers to overcome these biases.

In the international simulations, participants focused on core level aspects of situation awareness regardless of their levels of industry and OCC experience. Thus, they sought a level of awareness beyond basic aspects such as gaining a general overview. Rather, it appeared that these participants sought a level of awareness that enabled them to prepare for and identify the likely consequences of potential disruptions by identifying aspects relating to patterns and gaps in the flight display and maintenance and crewing limitations. In the domestic simulations, all participants focused on elementary aspects of situation awareness and the focus was most prominent for participants with extensive industry and OCC experience. This suggests that controllers in domestic OCCs may be satisfied with acquiring a sound awareness of fundamental information to prepare for disruptions. The findings are consistent with previous studies which have suggested the importance of experience for acquiring situation awareness. However, an unexpected result in the study was the focus by participants in the domestic simulations on elementary and core levels of situation awareness rather than advanced levels. This suggests that many participants limit their acquisition of situation awareness to the extent that they may fail to take into account more critical and complex aspects and raises an issue for OCC management to recognise possible deficiencies in the way that controllers develop situation awareness. Thus, it is important for OCC management to expand the repertoire of appropriate aspects of situation awareness at complex levels to enable controllers to develop their levels of situation awareness for managing disruptions.

The study also examined the relationship between situation awareness and expertise relying on previous research that suggested that decision-makers may be categorised as experts in their domain after ten years' experience. In the international simulations, participants focused on core level categories of situation awareness regardless of their level of expertise. However, participants with expertise in industry focused on more advanced level categories than novices. Further, participants with OCC domain expertise focused to a greater extent on advanced categories than all other participants. The conclusion drawn from these findings suggests that OCC expertise is likely to be most beneficial for assisting controllers to identify advanced categories. In other words, OCC expertise may be more relevant than industry expertise for acquiring situation awareness in international OCCs. This finding is significant in that advanced knowledge of situation awareness of experts could be drawn upon to assist in expanding the range and complexity of issues to be considered by novices. In the domestic simulations, all participants focused extensively on elementary aspects of situation awareness, suggesting that in the high levels of complexity characteristic of a domestic OCC environment, controllers may need high levels of expertise to enable them to develop a sound awareness of fundamental aspects such as maintenance, crewing, and short ground times which are critical to decision-making in this domain.

A number of conclusions can be drawn from these findings. First, expertise in both industry and OCC appears to be important for gaining situation awareness. Second, extensive domain expertise appears to be important to enable controllers to gain awareness of domain-specific aspects. These findings have implications for methods of recruiting and training controllers in OCCs.

Situation Awareness and Decision Considerations

During the simulations, participants took into account decision considerations prior to selecting decision alternatives. This step in the decision-making process is not well explained in the literature and is deficient in decision-making models. Further, it was evident that participants regarded some decision considerations as fundamental to a situation and other considerations as critical to the selection of decision alternatives. In view of this, the decision considerations were classified according to complexity in a similar way that the aspects of situation awareness were classified.

In the international simulations, participants focused predominantly on core level decision considerations in each of scenarios presented during the simulations. However, in more complex scenarios they also emphasised advanced level considerations. This was similar to the extent to which they emphasised core and advanced levels of situation awareness. In the domestic simulations, participants focused extensively on elementary and core decision considerations. These participants had emphasised elementary and core aspects of the flight display to gain situation awareness. The findings suggest that the level of situation awareness appears to influence the levels of decision considerations taken into account during decision-making. In other words, basic levels of situation awareness appear to lead only to basic levels of decision considerations. This means that for controllers to develop advanced decision considerations, they need to be encouraged to seek out advanced levels of situation awareness. The findings provide strong evidence to support the contention by Endsley, Bolté and Jones (2003) that situation awareness is a key factor in the decision-making process.

In terms of implications for OCCs, the findings suggest that developing a more enhanced level of situation awareness is likely to lead to decision considerations that best assist controllers to manage operational disruptions. In contrast, participants with inadequate situation awareness may only identify rudimentary decision considerations that may limit the decision-making process. This is an important finding for the study as it suggests that situation awareness, whether provided initially in a briefing or gained during a disruption, is absolutely critical for the optimum identification of potential problems and a range of viable solutions.

Expertise and Decision Considerations

The findings of the study indicate that participants' levels of expertise were critical to their decision-making processes during the simulations. This was demonstrated by the very different approaches novices and experts used during the simulations. In both the international and domestic simulations, novices appeared to have difficulty comprehending the disruptions presented in the scenarios and frequently requested further information to help clarify issues. However, they drew on this information in a haphazard way. They considered aspects that may address immediate problems but failed to consider further consequences. Thus, novices appeared to use a fragmented approach to decision-making and demonstrated their inability to gain an overall picture. These observations are consistent with previous studies examining novices and experts. In addition, novices often became overwhelmed with complexity and had difficulty coping with rapidly changing circumstances. As a result, they expressed concern for establishing what might work to resolve a situation.

In contrast, experts appeared to comprehend situations more readily and examined situations in far greater detail than novices. They were quick to clarify information and could build a comprehensive picture of a situation by drawing together the relevant components. Thus, experts appeared to use a more methodical approach than novices. As previous research has suggested, experts seemed to know what alternatives might work and what might not work in a situation. This was evident from the way in which they could project the consequences of a disruption in order to identify potential solutions, and devise creative strategies such as delaying decisions until a situation changed or further information became available.

The results highlight the different approaches taken by novices and experts, drawing particular attention to deficiencies in decision-making processes of novices. This raises questions as to the best ways to overcome these deficiencies, suggesting the need for some training programme drawing on the knowledge of expert controllers to expedite the development of novices. Given the importance of acquiring situation awareness, particular attention should be given to the way in which experts ascertain information and use it to build a comprehensive picture of a situation. A programme should also focus on ways in which experts take into account decision considerations, optimise time for decision-making, and discern workable from unworkable solutions. OCCs need to optimise contact between experts and novices to maximise the transfer of knowledge. A means to accomplish this may be through a 'buddy' or mentoring arrangement.

In the international simulations, novices and experts focused predominantly on core decision considerations. However, in highly complex disruptions it appeared that both novices and experts also focused on advanced decision considerations. Thus, it appears that participants in international OCCs may not necessarily rely on their levels of expertise to identify decision considerations. This may be due to the uniqueness of each situation, requiring a general ability to identify decision considerations rather than drawing on previous experience. Novices focused on decision considerations in all categories, similar to the extent of experts. This occurred in each of the international simulations and suggests that in this environment, appropriate outcomes may still be achieved despite the limited expertise of controllers, and regardless of the level of complexity. This was an unexpected result, as it was anticipated that novices may not have had adequate experience to identify decision considerations to the same extent as experts. One explanation for this may be that controllers have sufficient time in which to identify decision considerations extensively. Another reason may be that novices receive a degree of training for decision-making in international OCCs. While the study recognises that controllers in international and domestic OCCs are likely to have received various forms of training, the influence of training on decision-making processes was not examined in the study. However, further studies need to be conducted to examine the extent to which specific training could develop controllers' abilities to recognise decision considerations.

In the domestic simulations, novices and experts focused predominantly on elementary and core decision considerations. It was also evident that experts focused to a greater extent than novices on all categories of decision considerations. This suggests that expertise in industry and OCC influences decision-making processes in domestic OCCs to a greater extent than in international OCCs and highlights differences in levels of expertise required for decision-making in international and domestic OCCs. As a result, a gap is evident in the way that novices and experts identify decision considerations. This gap is likely to lead to a failure to identify sufficient and feasible decision alternatives from which decisions can be selected. One way of addressing this deficiency would be to develop a computer-based expert system that could generate various decision considerations according to any given operational disruption. This system could help to overcome the limitations of relying on controllers to generate considerations and could also help to overcome controllers' past industry experience acting as a possible bias. A further use of the system could complement existing OCC controller training by providing specific scenario-based disruptions to suit international and domestic OCCs. The scenarios could draw on previous disruptions to enable trainees to see the potential consequences of particular decision alternatives. This would enable trainees to practice disruption recovery strategies. In addition, training scenarios should be purposely designed for domestic OCCs that enable trainees to practice making decisions under time constraints.

In each of the scenarios in the international and domestic simulations, the advanced performance of experts compared with novices led to more comprehensive and workable decision alternatives upon which final decision choices could be relied. Thus, the study builds on previous work attesting to the superior performance of experts when faced with complex problems in the OCC domain.

Decision-making Styles and Decision Considerations

Rational Decision-making Style

It was evident that participants used a high rational DMS to identify decision considerations in both the international and domestic simulations. Participants with expertise in industry and OCC appeared to use a high rational DMS to a greater extent than novices. The use of high rational DMS was apparent to a greater extent in highly complex disruptions and occurred in both international and domestic environments. Previous research has contended that rational decision-making in uncertain, dynamic environments is unlikely to be successful based on the assumption that decision-makers need to process extensive information rapidly and assess the situations, often with limited time. In other words, decision-makers in these environments do not have sufficient time to adopt a step-by-step process prescribed by the rational decision-making style.

Thus, the finding that decision-makers use a high rational DMS regardless of the complexity of a problem was unexpected and contrary to previous research, raising an important point. While the use of high rational DMS in the international environment may be explained by the longer times generally available for decision-making, the same explanation does not account for the domestic environment where decision times may be restricted. This suggests that a rational decision-making style may enhance decision-making regardless of the complexity evident in a decision-making environment. Thus, an opportunity exists for OCCs to develop a specific training aid that incorporates the steps of a rational decision-making style. This supports the development of a computerised expert system that could mimic the steps involved in decision-making using a rational approach and would assist controllers' decision-making processes by guiding them through the decision process. The system would need to incorporate the enhancements suggested by the study such as the categories of situation awareness and decision considerations as this would assist controllers to identify crucial aspects relevant to the decision-making process.

Intuitive Decision-making Style

Comments were also classified according to participants' use of high intuitive decision-making styles. In situations of least complexity in both the international and domestic simulations, little use of intuitive DMS by either novices or experts was evident. However, in complex disruptions both novices and experts demonstrated considerable use of high intuitive DMS. It was also evident that experts appeared to use an intuitive approach more extensively than novices. In other words, decision-makers in OCCs appear to use an intuitive approach regardless of their level of expertise, but experts with experience in industry and OCC use an intuitive approach more than novices. Thus, the findings provide strong evidence of the use of intuition for decision-making in OCCs and build on extensive previous research contending that decision-makers are likely to draw on experience and use intuition for decision-making in dynamic, complex situations. The emphasis on high intuitive decision-making style suggests that developing participants' intuitive approach is likely to lead to improved decision-making processes. Thus, the development of a computerised expert system for training controllers also needs to incorporate a means for capturing ways in which experts use their intuition for decision-making. While some views contend that decision-making occurs in a continuum (from a rational approach to an intuitive one), the findings of the study do not support this. Rather, the findings suggest that controllers use a combination of both intuitive and rational styles of decision-making, in particular when managing complex problems.

Contributions and Implications

Situation Awareness

The study relied on Endsley's (1988, 1995a) definition of situation awareness and used her model to examine ways in which participants developed situation awareness. Although the first level of Endsley's model refers to the perception of elements in an environment, the model does not distinguish among these elements in terms of weighting them according to their relative importance. It was evident from the study that participants regarded some elements as fundamental and other elements as critical for the accumulation of situation awareness. In view of this, the elements of situation awareness were classified and weighted according to complexity. The study builds on Endsley's work by expanding the first level of her model which should help to improve understanding of ways in which decision-makers develop situation awareness and enable decision-makers to recognise differences between elements of situation awareness. The findings should lead to a greater understanding of situation awareness in other areas of aviation and domains such as emergency services, medicine, driving, war strategy, and power plant operations.

Further research could focus on the extent to which elements might be weighted in other domains as this is likely to lead to a greater understanding of the development of situation awareness in these domains. The findings of the study suggest a number of important research questions: (a) Can the most relevant elements of situation awareness be identified and weighted in particular domains? (b) In terms of Endsley's (1995a) model, to what extent does identifying the most relevant elements in a domain lead to greater levels of comprehension and projection of situation awareness in the decision-making process? (c) What is the relationship between situation awareness and information completeness? (d) How can information be tailored to specific situations to provide advanced levels of situation awareness? (e) What mechanisms can be put into place to assist decision-makers identify and prioritise relevant information to best suit a situation? (f) To what extent can improvements in situation awareness lead to identification of more advanced decision considerations?

Situation Awareness in Aviation

Very few studies have examined situation awareness in aviation other than with commercial or military pilots and air traffic controllers. This study has established that situation awareness is fundamental to decision-making processes of controllers in OCCs. This suggests that further research is needed to examine situation awareness in other aviation domains and raises a number of research questions: (a) To what extent is situation awareness critical in aviation domains such as ground handling, airport operations, load control, and crew rostering? (b) Can classification of aspects of situation awareness in other aviation domains lead to improvements in decision-making in those domains? (c) Can classification of aspects of situation awareness be incorporated into pilot and air traffic controller decision-making processes to improve situation awareness in these domains?

Situation Awareness in OCCs

The study demonstrated that situation awareness is vital for decision-making in OCCs. The study also established that acquiring situation awareness in OCCs is a cumulative process in which controllers gain an initial level of awareness in a familiarisation stage and enhance this initial awareness by gathering information during disruptions. Notably, the study concluded that situation awareness developed during the familiarisation stage may be more important than awareness gained during disruptions, in particular in domestic OCCs. This is due to the limited time to accumulate and analyse information. This enhanced awareness is likely to prepare controllers more thoroughly for disruptions in terms of assessing situations, understanding consequences that might occur and identifying solutions to problems. Further research needs to investigate training methods to develop situation awareness as this is likely to lead to improved decision-making processes. The need then, to improve OCC controllers' situation awareness raises several research questions: (a) To what extent can training influence ways in which controllers acquire initial situation awareness? (b) To what extent can information provide controllers with optimum levels of situation awareness? (c) What methods can be developed to ensure the accuracy of information for controllers? (d) In what ways can information be sorted and presented to controllers according to the nature of the problem? (e) Is there an optimum level of information that ensures controllers develop situation awareness and ensures that a high volume of information does not detract from awareness?

Although team situation awareness was not examined in the study, the increasing complexity of decision-making in OCCs could lead to further research to investigate the extent to which team situation awareness is likely to influence decision-making processes and prompts further research questions: (a) To what extent is team situation awareness evident in OCCs? (b) To what extent can team situation awareness lead to improved decision-making processes in OCCs?

Decision-making

A number of rational decision-making models have been outlined depicting decision-making occurring in a number of steps. Despite differences among the models, the fundamental stages common to all of them include problem identification, generation and assessment of alternatives, and selection of a course of action. An important finding in the study was the recognition that participants took several decision considerations into account prior to formulating the decision alternatives. Although the stage of recognising decision considerations may exist implicitly in decision-making models, the study demonstrated that controllers in OCCs clearly identify considerations as part of the decision-making process. Accordingly, this study draws attention to the importance of including a step in rational decision-making approaches models that explicitly depicts the identification and weighting of decision considerations. A similar step may clarify Klein's (1993) Recognition Primed Decision (RPD) as decision-makers mentally simulate possible actions to take.

Experience and Expertise

The results of the study emphasised the importance of both industry and OCC experience for developing situation awareness and the findings suggest that specific aviation industry experience is beneficial to decision-making processes in OCCs. Thus, further research could address several questions: (a) Is there an optimum length of service in the aviation industry that benefits decision-making processes in OCCs? (b) Is there an optimum diversity of experience in specific aviation areas that may best prepare controllers for working in OCCs? The findings of the study also challenge the definition of expertise being based on years of experience. An important conclusion of the study is that expertise in decision-making should be based more appropriately on the acquisition of certain decision-making sub-skills that enable decision-makers to identify and classify elements of situation awareness, identify and classify a range of decision considerations, and generate decision alternatives that result in optimum solutions to a problem. Thus, further research could address the following questions: (a) To what extent can a set of decision-making sub-skills of experts be identified? (b) To what extent can training aids facilitate a transfer of knowledge to enable novices to learn these sub-skills from experts?

Decision-making Styles

The study examined the use of rational and intuitive decision-making styles of controllers in OCCs. Past research suggests that rational decision-making styles are unlikely to be successful in complex, dynamic environments where time is limited to gather information and compare alternatives. However, it was evident in the study that participants used a high rational decision-making style in the most complex disruptions of the simulations. Past research also suggests that experience forms the basis of intuitive decision-making. However, in the study participants drew on intuition in highly complex situations regardless of their experience. The findings prompt a number of research questions that could be addressed: (a) To what extent can levels of rational and intuitive decision-making be identified in other domains? (b) To what extent does the use of a combination of high rational and high intuitive decision-making styles lead to effective decision-making?

Naturalistic Decision-making

The approach adopted in the study enabled the examination of individual decision-making processes by replicating an environment as close as possible to the controllers' workplace. Although NDM theory relies on research in real-world, natural contexts, the preliminary study demonstrated the difficulty of capturing thought processes of controllers in OCCs as they made decisions during operational disruptions. In OCCs, controllers do not have time during disruptions to discuss their thought processes or explain the rationale for taking certain actions. Therefore, the verbalisations permitted a meaningful insight into participants' thoughts and behaviours. Consequently, the use of simulations in conjunction with think-aloud protocol in the main study enabled the capture of rich, qualitative data that would otherwise have been impossible to acquire. The simulations also overcame the limitations of other data collection methods such as questionnaires which were ill-suited for in-depth examination of individuals' perspectives. Thus, the study demonstrates the application of these methods as a means of accessing the thought processes that underpin decision-making in very complex situations. Further research using this method would enable the collection of additional data which may assist airlines to examine further, decision-making processes of controllers in OCCs. Think-aloud protocol could be appropriate for studies of decision-making in other domains in which the collection of participants' thought processes could increase understanding of individual decision-making processes.

Further Studies in OCCs.

Past studies have concluded that expert systems have provided OCC controllers with limited decision support, but have failed to cope with multiple, simultaneous problems common to an OCC environment. Research on decision-making in OCCs has not taken into account the human involvement in the process even though disruption recovery in OCCs is largely a manual process relying on controllers' intuition and experience. The study has provided an insight into individual decision-making processes of controllers. The study needs to be replicated by examining decision-making processes of controllers in OCCs beyond the Asia-Pacific region in order to test the generalisability of the conclusions drawn in this study.

Human Resources Selection Practices

The study demonstrated the need to employ experienced controllers in OCCs and emphasised the importance of industry as well as domain experience. Therefore, airline human resource departments need to understand the suitability of industry backgrounds when recruiting OCC controllers. Further research would help to identify a range of suitable backgrounds both from within the aviation industry and from other industries. An understanding of the importance of experience gained from departments closely associated with OCCs is also necessary. Therefore, a recommendation from the study is the establishment of a pathway approach that requires short-term placements in these associated departments with the final objective of entering the OCC as a trainee controller. The findings of the study indicate that domain experience influenced decision-making in international and domestic OCCs. This suggests that HR personnel require an in-depth knowledge of the particular requirements of the OCC environment for which they are recruiting. In OCCs which manage both international and domestic schedules, further consideration needs to be given to determine the most beneficial way for a trainee to develop decision-making processes in either or both environments. Selection of controllers has often been haphazard in OCCs. The development of a computer-based simulation model that replicated a series of operational disruptions for testing and assessing decision-making processes of potential candidates would provide an additional means for selection.

Human Resources Training Practices

The study identified the need for training in three areas: developing situation awareness, identifying decision considerations, and enhancing expertise levels of novices. It has argued the importance of situation awareness for decision-making processes in OCCs. Therefore, it is vital that initial situation awareness is gained as thoroughly as possible to prepare controllers for potential disruptions. One way to assist novices may be to develop a computer-based expert system that could be used as a training aid by displaying real-time schedules appropriate to the OCC. This would enable novices to test the possible consequences of failing to take into consideration various aspects of a situation. The expert system could act as a prompt or checklist to expand the repertoire of considerations available in the situation. The method of classification and weighting used in the study may also assist novices to recognise elementary, core, and advanced level aspects of situation awareness. This system could measure the extent to which novices acquire initial situation awareness and should assist them to develop levels of situation awareness as a means for preparing for disruptions.

The study has also identified the importance of decision considerations in the decision-making process. Consequently, the same computer-based expert system could be used as a training aid to develop simulations using scenarios taken from real-life disruptions. The use of simulations could be very beneficial in terms of training novices to identify decision considerations prior to generating a range of decision alternatives. A training programme could also incorporate a 'buddy' system to mentor novices in situations of least complexity or when time may be too limited for extensive examination and discussion of decision-making processes.

The real benefits from developing such a training aid are many fold. It could provide controllers with a quick reference guide for identifying and classifying elements of situation awareness and for identifying and classifying decision considerations. It could assist OCC management in the selection of future controllers and the training of new and existing controllers.

Concluding Comment

Although previous studies have examined decision-making processes in many high profile areas of aviation such as piloting and air traffic control, decision-making in OCCs has largely been overlooked. The researcher's personal industry experience as an OCC controller provided the impetus for this book. The interest in examining and improving decision-making processes stemmed from first-hand knowledge of disruption management in the complex OCC environment. The research has also drawn on personal academic background experience in the field of HR with a particular interest in examining and writing case studies.

Previous research in the OCC domain may have been overlooked because in real life, access to the data has been considered too difficult to obtain. The use of simulations and think-aloud protocol has gone some way to overcome this research hurdle in this field.

The findings should be enlightening to airline management for improving selection and training methods of controllers in OCCs. In conclusion, the book should create the spark for further research as decision-making processes of controllers in OCCs warrant far more examination, given the importance of the role of the OCC to an airline.

References

Abdelghany, K.F., Abdelghany, A.F. and Ekollu, G. (2008). An integrated decision support tool for airlines schedule recovery during irregular operations. *European Journal of Operational Research*, 185, 825–48.

Abdelghany, K.F., Shah, S.S., Raina, S. and Abdelghany, A.F. (2004). A model for projecting flight delays during irregular operation conditions. *Journal of Air Transport Management*, 10, 385–94.

Adams, M.J., Tenney, Y.J. and Pew, R.W. (1995). Situation awareness and the cognitive management of complex systems. *Human Factors*, 37(1), 85–104.

Adelman, L., Tolcott, M.A. and Bresnick, T.A. (1993). Examining the effect of information order on expert judgment. *Organizational Behavior and Human Decision Processes*, 56, 348–69.

Agor, W.H. (1984). *Intuitive Management: Integrating Left and Right Brain Management Skills*. Englewood Cliffs. New Jersey: Prentice-Hall.

Agor, W.H. (1986). *The Logic of Intuitive Decision-making: A Research-based Approach for Top Management.* New York: Quorum Books.

Ahituv, N., Igbaria, M. and Sella, A. (1998). The effects of time pressure and completeness of information on decision-making. *Journal of Management Information Systems*, 15(2), 153–72.

Ahituv, N. and Neumann, S. (1987). Decision-making and the value of information. In: R. Galliers (ed.), *Information Analysis: Selected Readings*. USA: Addison Wesley Publishing Co.

Ariely, D. and Zakay, D. (2001). A timely account of the role of duration in decision-making. *Acta Psychologica*, 108(2), 187–207.

Bainbridge, L. (1999). Processes underlying human performance. In: D.J. Garland, J.A. Wise and V.D. Hopkin (eds), *Handbook of Aviation Human Factors*. Mahwah, NJ: Lawrence Erlbaum Associates.

Balch, R.S., Schrader, S.M. and Ruan, T. (2007). Collection, storage and application of human knowledge in expert systems development. *Expert Systems*, 24(5), 346–55.

Ball, M.O. (2003). Introduction to the special issue on aviation operations research: Commemorating 100 years of aviation. *Transportation Science*, 37(4), 366–7.

Barnhart, C., Belobaba, P. and Odoni, A.R. (2003). Applications of operations research in the air transport industry. *Transportation Science*, 37(4), 368–91.

Baron, J. (1994). *Thinking and Deciding* (2nd ed.). New York: Cambridge University Press.

Barthelemy, J.P., Bisdorff, R. and Coppin, G. (2002). Human centered processes and decision support systems. *European Journal of Operational Research*, 136(2), 233–52.

Bazargan, M. (2004). *Airline Operations and Scheduling*. Aldershot: Ashgate.

Beach, L.R. and Lipshitz, R. (1993). Why classical decision theory is an inappropriate standard for evaluating and aiding most human decision-making. In: G.A. Klein, J. Orasanu, R. Calderwood and C. Zsambok (eds), *Decision-Making in Action: Models and Methods*. Norwood, NJ: Ablex Publishing Corporation.

Beach, L.R. and Mitchell, T. (1978). A contingency model for the selection of decision strategies. *Academy of Management Review*, 3, 439–49.

Bedard, J. (1991). Expertise and its relation to audit decision quality. *Contemporary Accounting Research*, 8, 198–222.

Benner, L. (1975). D.E.C.I.D.E. in the hazardous materials emergencies. *Fire Journal*, 69(4), 13–18.

Benner, P. (1984). *From Novice to Expert: Excellence and Power in Clinical Nursing Practice*. Menlo Park, CA: Addison Wesley Publishing Company.

Berg, B.L. (1995). *Qualitative Research Methods for the Social Sciences* (2nd ed.). Needham Heights, MA: Allyn and Bacon.

Blandford, A. and Wong, B.L.W. (2004). Situation awareness in emergency medical dispatch. *International Journal of Human-computer Studies*, 61(4), 421–52.

Bonabeau, E. (2003). Don’t trust your gut. *Harvard Business Review*, 81(5), 116–23.

Brehmer, B. (1999). Reasonable decision-making in complex environments. In: H. Montgomery and P. Juslin (eds), *Judgment and Decision-making*. Mahwah, NJ: Lawrence Erlbaum.

Bryer, D.J. (2006). Influences that drive clinical decision-making among rheumatology nurses. Part 1: literature review. *Musculoskeletal Care*, 4(3), 130–39.

Buchanan, L. and O’Connell, A. (2006). A brief history of decision-making. *Harvard Business Review*, 84(1), 32–41.

Cannon-Bowers, J.A., Salas, E. and Pruitt, J.S. (1996). Establishing the boundaries of a paradigm for decision-making research. *Human Factors*, 38(2), 193–205.

Cao, J. and Kanafani, A. (2000). The value of runway time slots for airlines. *European Journal of Operational Research*, 126(3), 491–500.

Carr, M. and Wagner, C. (2002). A study of reasoning processes in software maintenance management. *Information Technology and Management*, 3, 181–203.

Cellier, J-M., Eyrolle, H. and Marine, C. (1997). Expertise in dynamic environments. *Ergonomics*, 40(1), 28–50.

Cesna, M. and Mosier, K. (2005). Using a prediction paradigm to compare levels of expertise and decision-making among critical care nurses. In: D.A. Montgomery, R. Lipshitz and B. Brehmer (eds), *How Professionals Make Decisions*. Mahwah, NJ: Lawrence Erlbaum Associates.

Chase, W. and Simon, H.A. (1973). Perceptions in chess. *Cognitive Psychology*, 4, 55–81.

Chi, M., Glaser, R. and Farr, M. (eds). (1988). *The Nature of Expertise*. New Jersey: Erlbaum.

Chu, P.C. and Spires, E.E. (2001). Does time constraint on users negate the efficacy of decision support systems? *Organizational Behavior and Human Decision Processes*, 85(2), 226–49.

Clarke, M.D. (1997). The airline schedule recovery problem (technical report). Cambridge, MA: Massachusetts Institute of Technology: Operations Research Centre.

Clausen, J., Larsen, A. and Larsen, J. (2005). *Disruption Management in the Airline Industry – Concepts, Models and Methods*. Lyngby: Informatics and Mathematical Modelling, Technical University of Denmark.

Cohn, A.M. and Barnhart, C. (2003). Improving crew scheduling by incorporating key maintenance routing decisions. *Operations Research*, 51(3), 387–96.

Corner, J., Buchanan, J. and Henig, M. (2001). Dynamic decision problem structuring. *Journal of Multi-criteria Decision Analysis*, 10(3), 129–41.

Crichton, M. and Flin, R. (2002). Command decision-making. In: R. Flin and K. Arbuthnot (eds), *Incident Command: Tales from the Hot Seat*. Aldershot: Ashgate.

Croft, D.G., Banbury, S.P., Butler, L.T. and Berry, D.C. (2004). The role of awareness in situation awareness. In: S. Banbury and S. Tremblay (eds), *A Cognitive Approach to Situation Awareness: Theory and Application*. Aldershot: Ashgate.

Currey, J. and Botti, M. (2003). Naturalistic decision-making: A model to overcome methodological challenges in the study of critical care nurses' decision-making about patients' hemodynamic status. *American Journal of Critical Care*, 12(3), 206–11.

Dane, E. and Pratt, M.G. (2007). Exploring intuition and its role in managerial decision-making. *Academy of Management Review*, 32(1), 33–54.

Dastani, M., Hulstijn, J. and van der Torre, L. (2005). How to decide what to do? *European Journal of Operational Research*, 160(3), 762–84.

de Groot, A. (1978). *Thought and Choice in Chess* (2nd ed.). The Hague: Mouton (Original work).

Dreyfus, H.L. and Dreyfus, S.E. (1986). *Mind Over Machine: The Power of Human Intuition and Expertise in the Era of the Computer*. New York: The Free Press.

Endsley, M.R. (1988). Design and evaluation for situation awareness enhancement. Paper presented at the 32nd annual meeting of the Human Factors Society.

Endsley, M.R. (1995a). Toward a theory of situation awareness in dynamic systems. *Human Factors*, 37(1), 32–64.

Endsley, M.R. (1995b). Measurement of situational awareness in dynamic systems. *Human Factors*, 37(1), 65–84.

Endsley, M.R., Bolté, B. and Jones, D.G. (2003). *Designing for Situation Awareness: An Approach to User-centered Design*. London: Taylor and Francis.

Endvick, B. (1996). Strategic decision-making processes in opportunistic situations: Are entrepreneurs more comprehensive than managers? Paper presented at the Academy of Entrepreneurship, 2.

Ericsson, K.A. (2005a). Recent advances in expertise research: A commentary on the contributions to the special issue. *Applied Cognitive Psychology*, 19(2), 233–41.

Ericsson, K.A. (2005b). Superior decision-making as an integral quality of expert performance: Insights into the mediating mechanisms and their acquisition through deliberate practice. In: D.A. Montgomery, R. Lipshitz and B. Brehmer (eds), *How Professionals Make Decisions*. Mahwah, NJ: Lawrence Erlbaum Associates.

Ericsson, K.A. and Lehmann, A. (1996). Expert and exceptional performance: evidence of maximal adaption to task constraints. *Annual Review of Psychology, Annual*, 47, 273–306.

Ericsson, K.A. and Simon, H.A. (1993). *Protocol Analysis: Verbal Reports as Data* (Revised ed.). Cambridge, MA: MIT Press.

Flach, J.M. and Rasmussen, J. (2000). Cognitive Engineering: Designing for situation awareness. In: N.B. Sarter and R. Amalberti (eds), *Cognitive Engineering in the Aviation Domain*. Mahwah, NJ: Lawrence Erlbaum Associates.

Flin, R., Salas, E., Strub, M. and Martin, L. (eds). (1997). *Decision-making Under Stress: Emerging Themes and Applications*. Aldershot: Ashgate.

Flin, R., Stewart, K. and Slaven, G. (1996). Emergency decision-making in the offshore oil and gas industry. *Human Factors*, 38(2), 262–77.

Garland, D.J., Stein, E.S. and Muller, J.K. (1999). Air Traffic Controller memory: capabilities, limitations, and volatility. In: D.J. Garland, J.A. Wise and V.D. Hopkin (eds), *Handbook of Aviation Human Factors*. Mahwah, NJ: Lawrence Erlbaum Associates.

Gobet, F. and Simon, H.A. (2000). Five seconds or sixty? Presentation time in expert memory. *Cognitive Science*, 24(4), 651–82.

Goodwin, L.D. and Goodwin, W.L. (1985). Statistical techniques in AERJ articles, 1979–1983: the preparation of graduate students to read the educational research literature. *Educational Researcher*, 14(2), 5–11.

Haider, H. and Frensch, P.A. (1999). Information reduction during skill acquisition: The influence of task instruction. *Journal of Experimental Psychology: Applied*, 5(2), 129–51.

Hammond, K.R. (1993). Naturalistic decision-making from a Brunswikian viewpoint: its past, present, future. In: G. Klein, J. Orasanu, R. Calderwood and C. Zsambok (eds), *Decision-making in Action: Models and Methods*. Norwood, NJ: Ablex.

Hauss, Y. and Eyferth, K. (2003). Securing future ATM-concepts safety by measuring situation awareness in ATC. *Aerospace Science and Technology*, 7, 417–427.

Hayes-Roth, F., Waterman, D.A. and Lenat, D.B. (1983). *Building Expert Systems*. Reading, MA: Addison-Wesley.

Hedlund, J. and Sternberg, R.J. (2000). Practical intelligence: Implications for human resources research. *Research in Personnel and Human Resources Management*, 19, 1–52.

Helton, W.S. (2004). The development of expertise: Animal models? *The Journal of General Psychology*, 131(1), 86–96.

Hershey, D.A. and Walsh, D.A. (2000). Knowledge versus experience in financial problem solving performance. *Current Psychology*, 19(4), 261–91.

Hoffman, R.R., Shadbolt, N.R., Burton, A.M. and Klein, G. (1995). Eliciting knowledge from experts: A methodological analysis. *Organizational Behavior and Human Decision Processes*, 62(2), 129–58.

Hörmann, H-J. (1995). FOR-DEC: a prescriptive model for aeronautical decision-making. In: R. Fuller, N. Johnston and N. McDonald (eds), *Human Factors in Aviation Operations*. Aldershot: Ashgate.

Itoh, M. and Inagaki, T. (2004). A microworld approach to identifying issues of human-automation systems design for supporting operator's situation awareness. *International Journal of Human-computer Interaction*, 17(1), 3–24.

Janic, M. (2005). Modeling the large scale disruptions of an airline network. *Journal of Transportation Engineering*, 131(4), 249–60.

Janis, I. (1989). *Crucial Decisions: Leadership in Policymaking and Crisis Management*. New York: The Free Press.

Janis, I. and Mann, L. (1977). *Decision-making: A Psychological Analysis of Conflict, Choice, and Commitment*. New York: The Free Press.

Janney, J.J. and Dess, G.G. (2004). Can real-options analysis improve decision-making? Promises and pitfalls. *Academy of Management Executive*, 18(4), 60–75.

Jentsch, F., Barnett, J., Bowers, C. and Salas, E. (1999). Who is flying this plane anyway? What mishaps tell us about crew member role assignment and air crew situation awareness. *Human Factors*, 41(1), 1–13.

Jentsch, F., Bowers, B. and Salas, E. (2001). What determines whether observers recognize targeted behaviors in modeling displays? *Human Factors*, 43(3), 496–507.

Johnson, E.J., Payne, J.W. and Bettman, J.R. (1993). Adapting to time constraints. In: O. Svenson and A.J. Maule (eds), *Time Pressure and Stress in Human Judgments and Decision-making*. New York: Plenum Press.

Johnson, P.E., Zualkernan, I.A. and Tukey, D. (1993). Types of expertise: an invariant of problem solving. *International Journal of Man-Machine Studies*, 39, 641–65.

Kaber, D.B., Perry, C.M., Segall, N., McClernon, C.K. and Prinzel III, L.J. (2006). Situation awareness implications of adaptive automation for information processing in an air traffic control-related task. *International Journal of Industrial Ergonomics*, 36, 447–62.

Kaempf, G.L., Klein, G., Thorsden, M.L. and Wolf, S. (1996). Decision-making in complex naval command-and-control environments. *Human Factors*, 38(2), 220–31.

Kahneman, D. and Tversky, A. (1982). On the study of statistical intuitions. *Cognition*, 11(2), 123–41.

Kass, S.J., Cole, K.S. and Stanny, C.J. (2007). Effects of distraction and experience on situation awareness and simulated driving. *Transportation Research Part F: Traffic psychology and behavior*, 10(4), 321–9.

Khatrih, N. and Ng, A. (2000). The role of intuition in strategic decision-making. *Human Relations*, 53(1), 57–73.

Klahr, D. and Simon, H.A. (2001). What have psychologists (and others) discovered about the process of scientific discovery? *Current Directions in Psychological Science*, 10(3), 75–9.

Klein, G. (1993). A Recognition-primed decision (RPD) model of rapid decision-making. In: G. Klein, J. Orasanu, R. Calderwood and C. Zsambok (eds), *Decision-making in Action: Models and Methods*. Norwood, NJ: Ablex.

Klein, G. (1997). The current status of the naturalistic decision-making framework. In: R. Flin, E. Salas, M. Strub and L. Martin (eds), *Decision-making Under Stress: Emerging Themes and Applications*. Aldershot: Ashgate.

Klein, G. (1998). *Sources of Power: How People Make Decisions*. Cambridge: MIT Press.

Klein, G. (2001). Understanding and supporting decision-making. *Information, Knowledge, Systems Management*, 2, 291–6.

Klein, G. (2003). *The Power of Intuition*. New York: Doubleday.

Klein, G. and Klinger, D. (1991). Naturalistic decision-making. *Crew system ergonomics information center (CSERIAC) Gateway*, 2(1), 1–4.

Klein, G. and Weick, K.E. (2000). Decisions. *Across the Board*, 37(6), 16–22.

Kleindorfer, P.R., Kunreuther, H.C. and Schoemaker, P.J.H. (1993). *Decision Sciences: An Integrative Perspective*. New York: Cambridge University Press.

Kohl, N., Larsen, A., Larsen, J., Ross, A. and Tiourine, S. (2007). Airline disruption management – Perspectives, experiences and outlook. *Journal of Air Transport Management*, 13, 149–62.

Lederer, P.J. and Nambimadom, R.S. (1998). Airline Network Design. *Operations Research*, 46(6), 785–804.

Lettovsky, L. (1997). *Airline Operations Recovery: An Optimization Approach. Ph.D. Thesis*. Georgia Institute of Technology, Atlanta, GA.

Lettovsky, L., Johnson, E.L. and Nemhauser, G.L. (2000). Airline crew recovery. *Transportation Science*, 34(4), 337–48.

Leybourne, S. and Sadler-Smith, E. (2006). The role of intuition and improvisation in project management. *International Journal of Project Management*, 24, 483–92.

Lipshitz, R. (1993). Converging themes in the study of decision-making in realistic settings. In: G. Klein, J. Orasanu, R. Calderwood and C. Zsambok (eds), *Decision-making in Action: Models and Methods*. Norwood, NJ: Ablex.

Lohatepanont, M. and Barnhart, C. (2004). Airline schedule planning: Integrated models and algorithms for schedule design and fleet assignment. *Transportation Science*, 38(1), 19–32.

Lord, R.G. and Maher, K.J. (1993). *Leadership and Information Processing: Linking Perceptions and Performance*. New York: Routledge.

Marsh, B., Todd, P.M. and Gigerenzer, G. (2004). Cognitive heuristics: Reasoning the fast and frugal way. In: J.P. Leighton and R.J. Sternberg (eds), *The Nature of Reasoning*. New York: Cambridge University Press.

Mather, J. (1989). Beyond CRM to decisional heuristics: An airline generated model to examine accidents and incidents caused by crew errors in deciding. Paper presented at the Fifth International Symposium on Aviation Psychology, Ohio State University.

Matthews, M.D., Strater, L.D. and Endsley, M.R. (2004). Situation awareness requirements for infantry platoon leaders. *Military Psychology*, 16(3), 149–61.

Maule, A.J., Hockey, G.R.J. and Bdzola, L. (2000). Effects of time-pressure on decision-making under uncertainty: changes in affective state and information processing strategy. *Acta Psychologica*, 104(3), 283–301.

McLucas, A.C. (2003). *Decision-making: Risk Management, Systems Thinking and Situation Awareness*. Canberra: Argos Press.

Means, B., Salas, E., Crandall, B. and Jacobs, T. (1993). Training decision-makers for the real world. In: G. Klein, J. Orasanu, R. Calderwood and C. Zsambok (eds), *Decision-making in Action: Models and Methods*. Norwood NJ: Ablex.

Miles, M.B. and Huberman, A.M. (1994). *Qualitative Data Analysis* (2nd ed.). Thousand Oaks: Sage Publications Inc.

Niessen, C. and Eyferth, K. (2001). A model of the air traffic controller's picture. *Safety Science*, 37(2–3), 187–202.

Nutt, P. (1989). *Making Tough Decisions*. San Francisco, CA: Jossey-Bass.

Nutt, P. (1999). Surprising but true: Half the decisions in organizations fail. *Academy of Management Executive*, 13(4), 75–90.

O'Brien, K.S. and O'Hare, D. (2007). Situational awareness ability and cognitive skills training in a complex real-world task. *Ergonomics*, 50(7), 1064–91.

O'Hare, D. (1997). Cognitive ability determinants of elite pilot performance. *Human Factors*, 39(4), 540–52.

Orasanu, J. and Connolly, T. (1993). The reinvention of decision-making. In: G. Klein, J. Orasanu, R. Calderwood and C. Zsambok (eds), *Decision-making in Action: Models and Methods*. Norwood, NJ: Ablex.

Ordonez, L. and Benson, L. (1997). Decisions under time pressure: How time constraint affects risky decision-making. *Organizational Behavior and Human Decision Processes*, 71(2), 121–40.

Payne, J.W., Bettman, J.R. and Johnson, E.J. (1990). The adaptive decision-maker: Effort and accuracy in choice. In: R.M. Hogarth (ed.), *Insights in Decision-making*. Chicago: The University of Chicago Press.

Prince, C., Ellis, E., Brannick, M.T. and Salas, E. (2007). Measurement of team situation awareness in low experience level aviators. *International Journal of Aviation Psychology 2007*, 17(1), 41–57.

Rapajic, J. (2009). *Beyond Airline Disruptions*. Aldershot: Ashgate.

Rastegary, H. and Landy, F.J. (1993). The interactions among time urgency, uncertainty, and time pressure. In: O. Svenson and A.J. Maule (eds), *Time Pressure and Stress in Human Judgment and Decision-making*. New York: Plenum Press.

Roberts, M.J. (2004). Heuristics and reasoning 1. In: J.P. Leighton and R.J. Sternberg (eds), *The Nature of Reasoning*. New York: Cambridge University Press.

Robertson, M.M. and Endsley, M.R. (1995). The role of crew resource management (CRM) in achieving team situation awareness in aviation settings. In: R. Fuller, N. Johnston and N. McDonald (eds), *Human Factors in Aviation Operations*. Aldershot: Ashgate.

Rosenberger, J.M., Johnson, E.L. and Nemhauser, G.L. (2003). Rerouting aircraft for airline recovery. *Transportation Science*, 37(4), 408–21.

Rosenberger, J.M., Schaefer, A.J., Goldsman, D., Johnson, E.L., Kleywegt, A.J. and Nemhauser, G.L. (2002). A stochastic model of airline operations. *Transportation Science*, 36(4), 357–77.

Russo, J.A. (2006). Experience, learning, and the process of expert development. *International Advances in Economic Research*, 12, 261–75.

Sadler-Smith, E. and Shefy, E. (2004a). The intuitive executive: Understanding and applying 'gut feel' in decision-making. *Academy of Management Executive*, 18(4), 76–91.

Sadler-Smith, E. and Shefy, E. (2004b). Professional standards research: developing intuition. *People Management,* 51–2.

Sadler-Smith, E. and Shefy, E. (2007). Developing intuitive awareness in management education. *Academy of Management Learning and Education*, 6(2), 186–205.

Salas, E., Prince, C., Baker, D.P. and Shrestha, L. (1995). Situation awareness in team performance: implications for measurement and training. *Human Factors*, 37(1), 123–36.

Sarter, N.B. and Woods, D.D. (1991). Situation awareness: A critical but ill-defined phenomenon. *The International Journal of Aviation Psychology*, 1(1), 45–57.

Scott, S. and Bruce, R. (1995). Decision-making style: The development and assessment of a new measure. *Educational and Psychological Measurement*, 55(5), 818–31.

Selnes, F. and Troye, S.V. (1989). Buying expertise, information search and problem solving. *Journal of Economic Psychology*, 10, 411–28.

Shafir, E., Simonson, I. and Tversky, A. (1997). Reason-based choice. In D.K. Goldstein and R.M. Hogarth (eds), *Research on Judgment and decision-making: Currents, Connections and Controversies*. New York: Cambridge University Press.

Shanteau, J., Weiss, D.J., Thomas, R.P. and Pounds, J.C. (2002). Performance-based assessment of expertise: How to decide if someone is an expert or not. *European Journal of Operational Research*, 136(2), 253–63.

Simon, H.A. (1976). *Administrative Behavior: A Study of Decision-making Process in Administrative Organizations* (4th ed.). New York: Free Press.

Simon, H.A. (1982). *Models of Bounded Rationality*. Cambridge, MA: MIT Press.

Simon, H.A. (1986). Alternative visions of rationality. In: H.R. Arkes and J.S. Hammond (eds), *Judgment and Decision-making: An Interdisciplinary Reader*. Cambridge: University of Cambridge.

Simon, H.A. (1987). Making management decisions: The role of intuition and emotion. *The Academy of Management Executive*, 1, 57–64.

Simon, H.A. and Chase, W. (1973). Skill in chess. *American Scientist*, 61, 394–403.

Sohn, Y.W. and Doane, S.M. (2004). Memory processes of flight situation awareness: interactive roles of working memory capacity, long-term working memory, and expertise. *Human Factors*, 46(3), 461–75.

Sonenshein, S. (2007). The role of construction, intuition, and justification in responding to ethical issues at work: The sensemaking-intuition model. *Academy of Management Review*, 32(4), 1022–40.

Sonnenwald, D.H., Maglaughlin, K.L. and Whitton, M.C. (2004). Designing to support situation awareness across distances: an example from a scientific collaboratory. *Information Processing and Management*, 40, 989–1011.

Soumis, F., Ferland, J.A. and Rousseau, J-M. (1980). A model for large-scale aircraft routing and scheduling problems. *Transportation Research Part B*, 14, 191–201.

Sriram, C. and Haghani, A. (2003). An optimization model for aircraft maintenance scheduling and re-assignment. *Transportation Research Part A*, 37, 29–48.

Stewart, T.A. (2006). Did you ever have to make up your mind? *Harvard Business Review*, 84(1), 12.

Talluri, K.T. (1996). Swapping applications in a daily airline fleet assignment. *Transportation Science*, 30(3), 237–44.

Todd, P.M. and Gigerenzer, G. (2000). Précis of simple heuristics that make us smart. *Behavioral and Brain Sciences*, 23, 727–80.

Tversky, A. and Kahneman, D. (1982). Judgment under uncertainty: Heuristics and biases. In: D. Kahneman, P. Slovic and A. Tversky (eds), *Judgment Under Uncertainty: Heuristics and Biases*. New York: Cambridge University Press.

Tversky, A. and Kahneman, D. (1985). The framing of decisions and the psychology of choice. In: G. Wright (ed.), *Behavioral Decision-making*. New York: Plenum Press.

van den Bosch, K. and Helsdingen, A.S. (2002). Improving tactical decision-making through critical thinking. Paper presented at the Human Factors and Ergonomics Society, Baltimore, MA.

Vaughan, F.E. (1979). *Awakening Intuition*. New York: Anchor Books.

Watson, S. (1992). The presumptions of prescription. *Acta Psychologica*, 80(1–3), 7–31.

Weiss, D.J. and Shanteau, J. (2003). Empirical assessment of expertise. *Human Factors*, 45(1), 104–16.

Wiegman, D.A., Goh, J. and O'Hare, D. (2002). The role of situation assessment and flight experience in pilots' decisions to continue visual flight rules flight into adverse weather. *Human Factors*, 44(2), 189–97.

Wong, B.L.W. (2000). The integrated decision model in emergency dispatch management and its implications for design. *Australian Journal of Information Systems*, 2(7), 96–101.

Wu, C.L. (2010). *Airline Operations and Delay Management: Insights from Airline Economics, Networks and Strategic Schedule Planning*, Aldershot: Ashgate.

Yan, S. and Tu, Y. (1997). Multifleet routing and multistop flight scheduling for schedule perturbation. *European Journal of Operational Research*, 103, 155–69.

Yan, S. and Yang, D.H. (1996). A decision support framework for handling schedule perturbation. *Transportations Research,* B30(6), 405–19.

Yates, J. (1993). The opportunity of qualitative research. *The Journal of Business Communication*, 30(2), 199–200.

Yates, J.F., Veinott, E.S. and Patalano, A.L. (2003). Hard decision, bad decisions: On decision quality and decision aiding. In: S. Schneider and J. Shanteau (eds), *Emerging Perspectives on Judgment and Decision Research.* New York: Cambridge University Press.

Appendix A

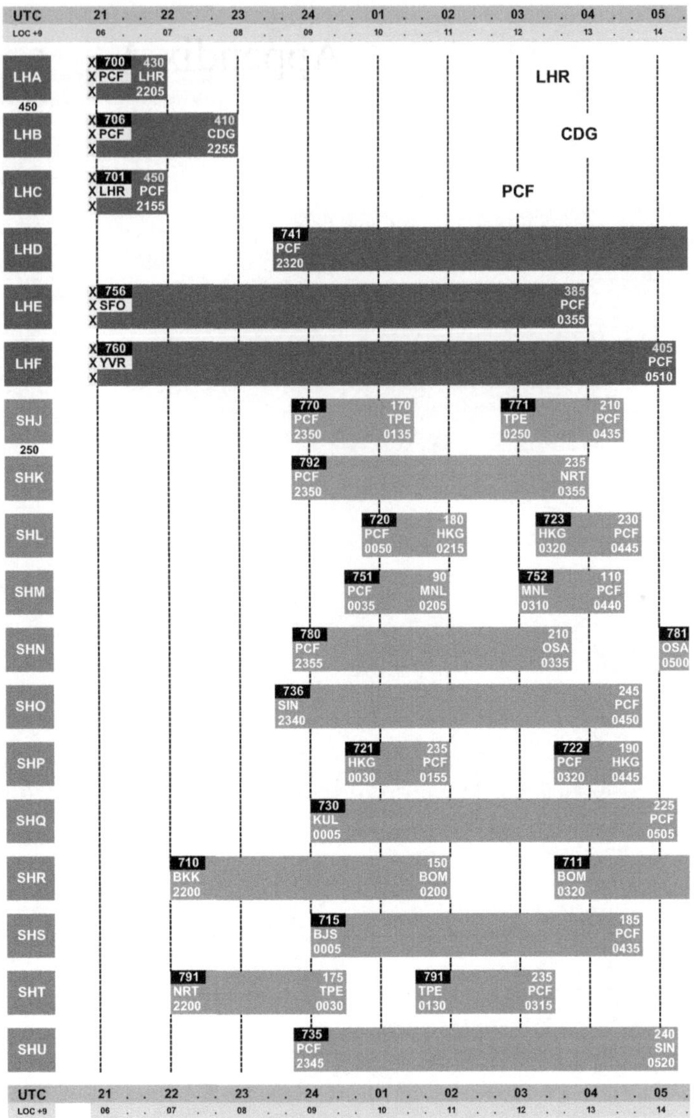

Figure A.1 International simulations (day 1)

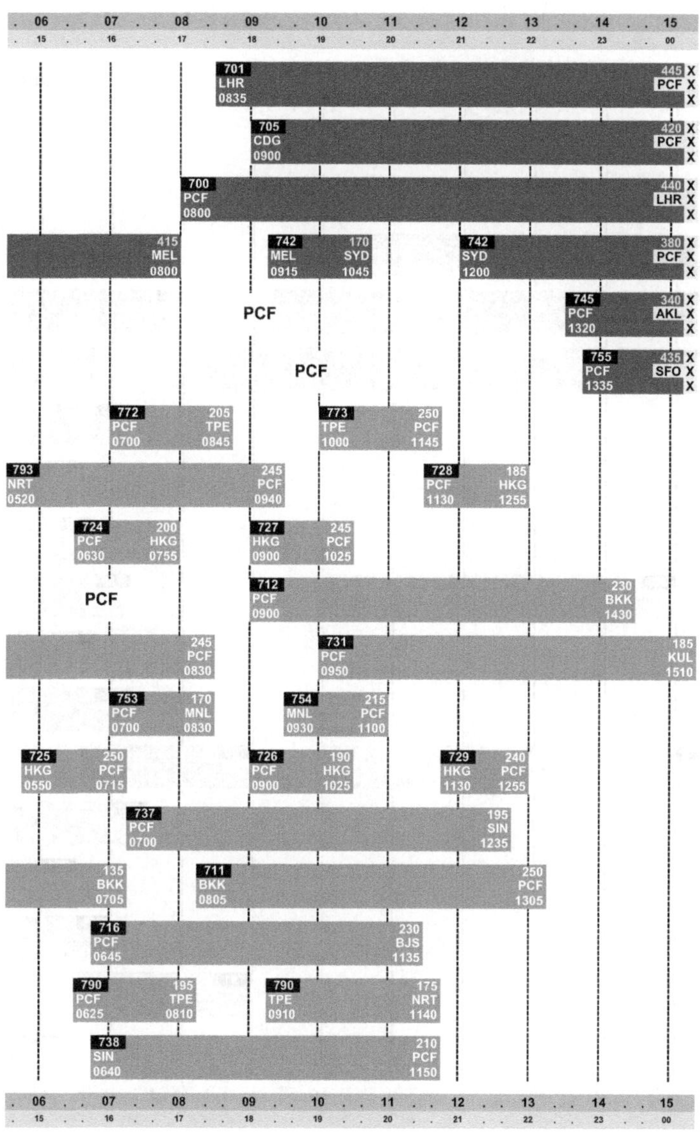
. 06 . . 07 . . 08 . . 09 . . 10 . . 11 . . 12 . . 13 . . 14 . . 15
. 15 . . 16 . . 17 . . 18 . . 19 . . 20 . . 21 . . 22 . . 23 . . 00
701 LHR 0835
445 PCF X X X
705 CDG 0900
420 PCF X X X
700 PCF 0800
440 LHR X X X
415 MEL 0800
742 MEL 0915
170 SYD 1045
742 SYD 1200
380 PCF X X X
745 PCF 1320
340 AKL X X X
PCF
755 PCF 1335
435 SFO X X X
PCF
772 PCF 0700
205 TPE 0845
773 TPE 1000
250 PCF 1145
793 NRT 0520
245 PCF 0940
728 PCF 1130
185 HKG 1255
724 PCF 0630
200 HKG 0755
727 HKG 0900
245 PCF 1025
712 PCF 0900
230 BKK 1430
PCF
245 PCF 0830
731 PCF 0950
185 KUL 1510
753 PCF 0700
170 MNL 0830
754 MNL 0930
215 PCF 1100
725 HKG 0550
250 PCF 0715
726 PCF 0900
190 HKG 1025
729 HKG 1130
240 PCF 1255
737 PCF 0700
195 SIN 1235
135 BKK 0705
711 BKK 0805
250 PCF 1305
716 PCF 0645
230 BJS 1135
790 PCF 0625
195 TPE 0810
790 TPE 0910
175 NRT 1140
738 SIN 0640
210 PCF 1150
. 06 . . 07 . . 08 . . 09 . . 10 . . 11 . . 12 . . 13 . . 14 . . 15
. 15 . . 16 . . 17 . . 18 . . 19 . . 20 . . 21 . . 22 . . 23 . . 00

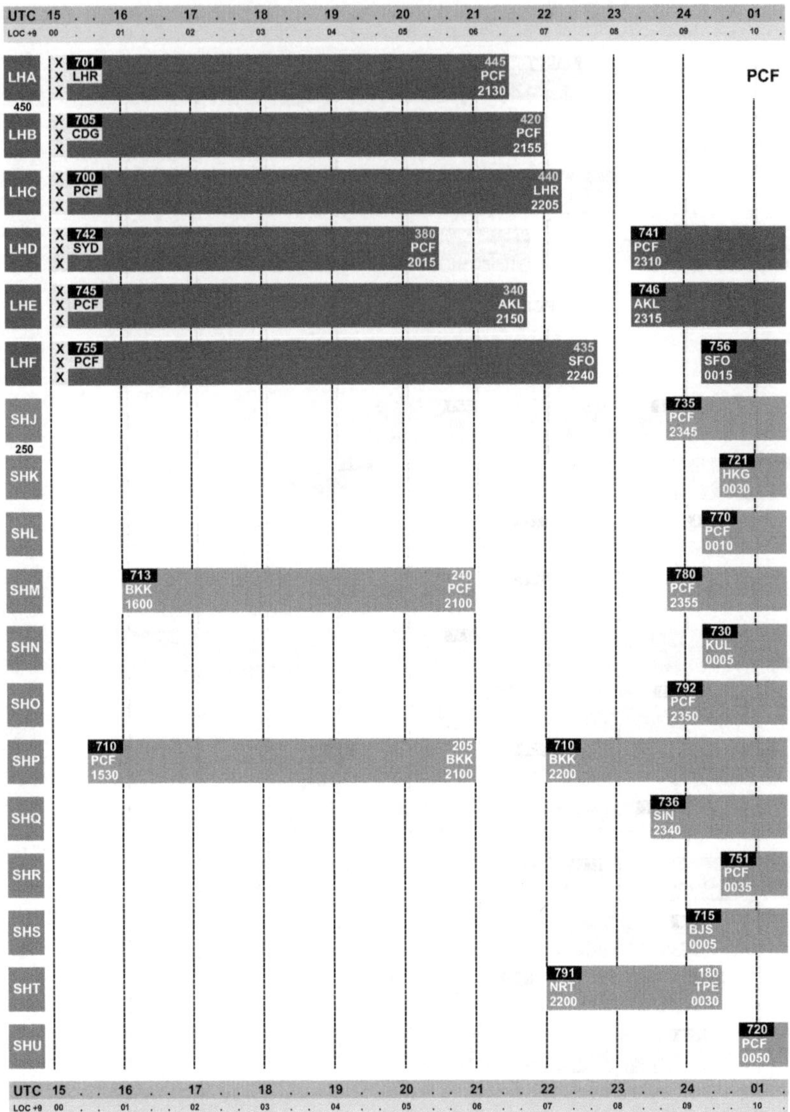

Figure A.2 International simulations (day 2)

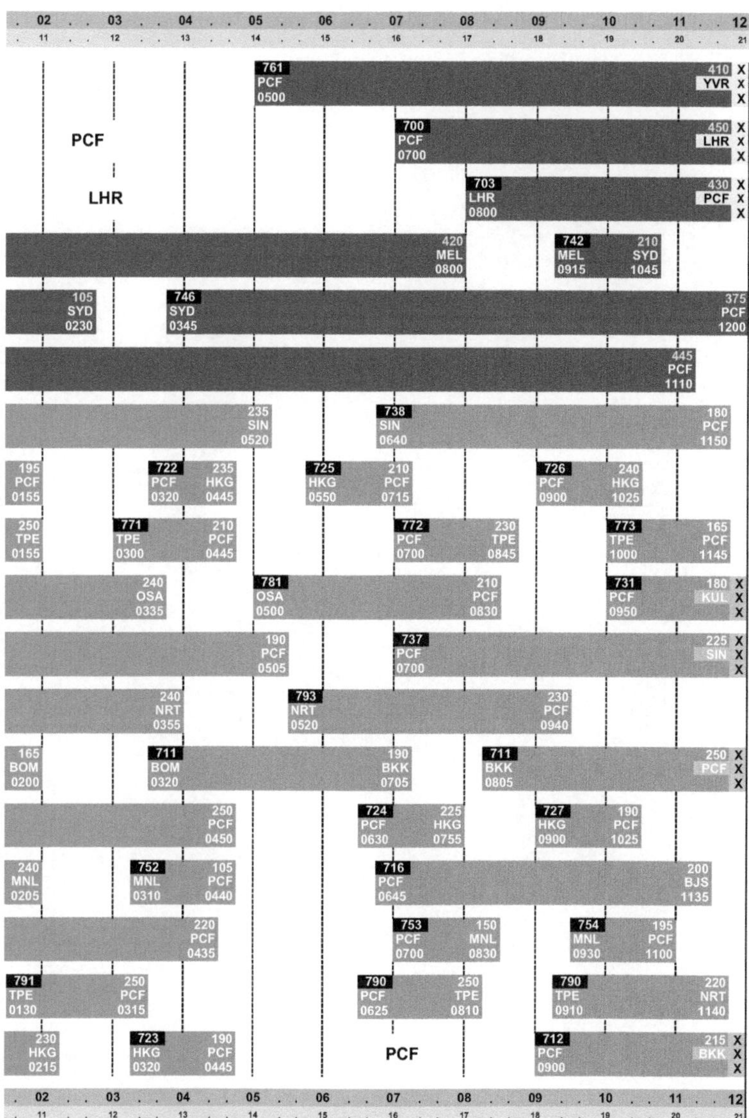

02 03 04 05 06 07 08 09 10 11 12
11 12 13 14 15 16 17 18 19 20 21
PCF
LHR
761 PCF 0500
410 YVR
700 PCF 0700
450 LHR
703 LHR 0800
430 PCF
420 MEL 0800
742 MEL 0915
210 SYD 1045
105 SYD 0230
746 SYD 0345
375 PCF 1200
445 PCF 1110
235 SIN 0520
738 SIN 0640
180 PCF 1150
195 PCF 0155
722 PCF 0320
235 HKG 0445
725 HKG 0550
210 PCF 0715
726 PCF 0900
240 HKG 1025
250 TPE 0155
771 TPE 0300
210 PCF 0445
772 PCF 0700
230 TPE 0845
773 TPE 1000
165 PCF 1145
240 OSA 0335
781 OSA 0500
210 PCF 0830
731 PCF 0950
180 KUL
190 PCF 0505
737 PCF 0700
225 SIN
240 NRT 0355
793 NRT 0520
230 PCF 0940
165 BOM 0200
711 BOM 0320
190 BKK 0705
711 BKK 0805
250 PCF
250 PCF 0450
724 PCF 0630
225 HKG 0755
727 HKG 0900
190 PCF 1025
240 MNL 0205
752 MNL 0310
105 PCF 0440
716 PCF 0645
200 BJS 1135
220 PCF 0435
753 PCF 0700
150 MNL 0830
754 MNL 0930
195 PCF 1100
791 TPE 0130
250 PCF 0315
790 PCF 0625
250 TPE 0810
790 TPE 0910
220 NRT 1140
230 HKG 0215
723 HKG 0320
190 PCF 0445
PCF
712 PCF 0900
215 BKK

Appendix B

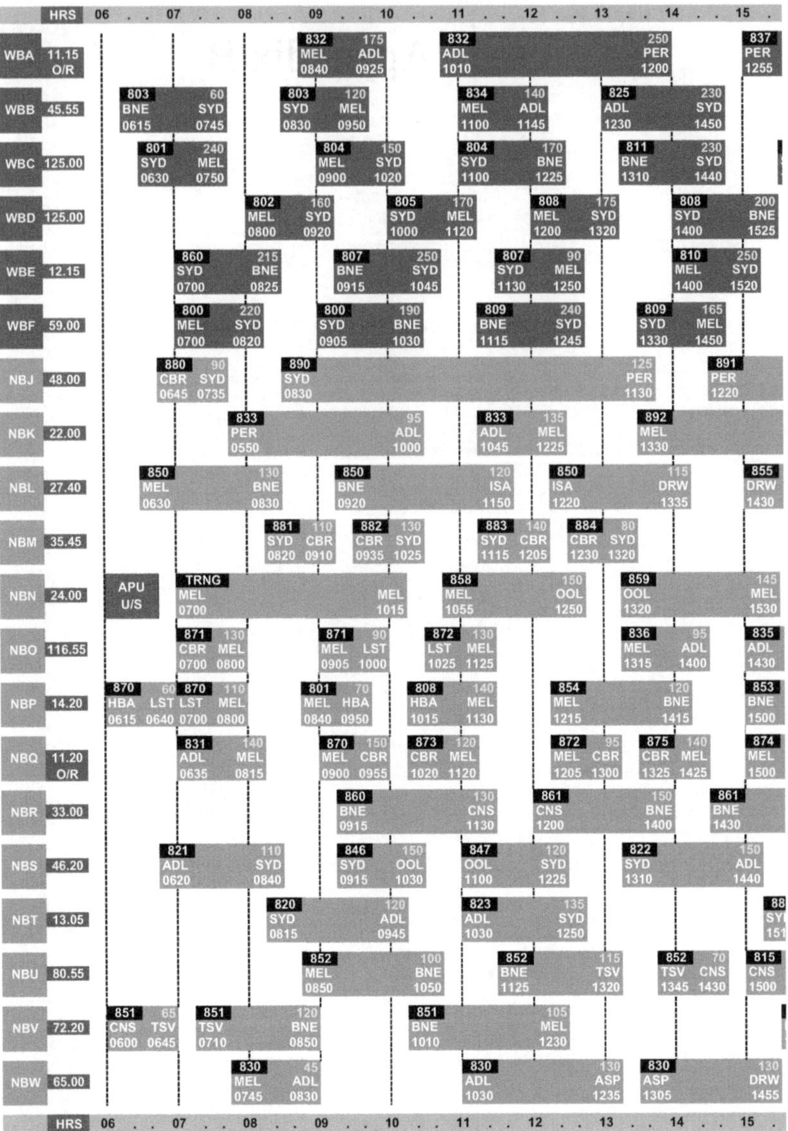

Figure B.1 Domestic utilisation (day 1)

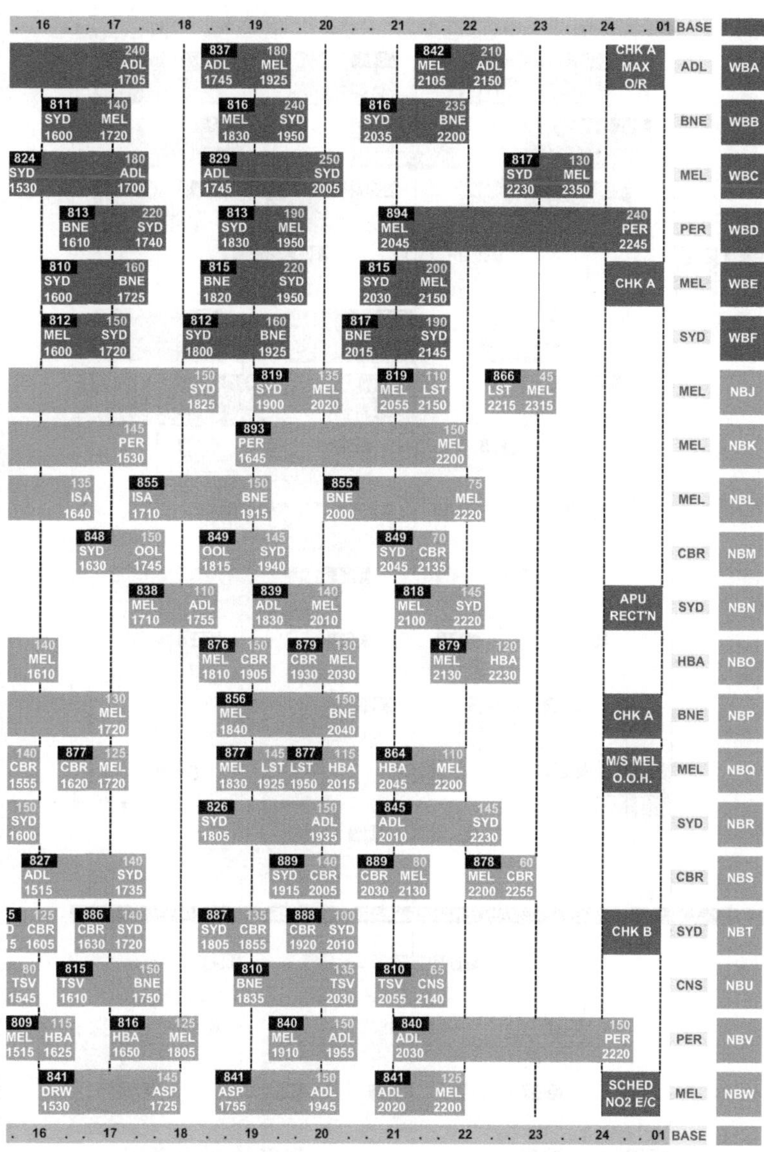
. 16 . . 17 . . 18 . . 19 . . 20 . . 21 . . 22 . . 23 . . 24 . . 01 BASE
240 ADL 1705 | 837 180 ADL MEL 1745 1925 | 842 210 MEL ADL 2105 2150 | CHK A MAX O/R | ADL | WBA
811 140 SYD MEL 1600 1720 | 816 240 MEL SYD 1830 1950 | 816 235 SYD BNE 2035 2200 | BNE | WBB
824 180 SYD ADL 1530 1700 | 829 250 ADL SYD 1745 2005 | 817 130 SYD MEL 2230 2350 | MEL | WBC
813 220 BNE SYD 1610 1740 | 813 190 SYD MEL 1830 1950 | 894 240 MEL PER 2045 2245 | PER | WBD
810 160 SYD BNE 1600 1725 | 815 220 BNE SYD 1820 1950 | 815 200 SYD MEL 2030 2150 | CHK A | MEL | WBE
812 150 MEL SYD 1600 1720 | 812 160 SYD BNE 1800 1925 | 817 190 BNE SYD 2015 2145 | SYD | WBF
150 SYD 1825 | 819 135 SYD MEL 1900 2020 | 819 110 MEL LST 2055 2150 | 866 45 LST MEL 2215 2315 | MEL | NBJ
145 PER 1530 | 893 150 PER MEL 1645 2200 | MEL | NBK
135 ISA 1640 | 855 150 ISA BNE 1710 1915 | 855 75 BNE MEL 2000 2220 | MEL | NBL
848 150 SYD OOL 1630 1745 | 849 145 OOL SYD 1815 1940 | 849 70 SYD CBR 2045 2135 | CBR | NBM
838 110 MEL ADL 1710 1755 | 839 140 ADL MEL 1830 2010 | 818 145 MEL SYD 2100 2220 | APU RECT'N | SYD | NBN
140 MEL 1610 | 876 150 MEL CBR 1810 1905 | 879 130 CBR MEL 1930 2030 | 879 120 MEL HBA 2130 2230 | HBA | NBO
130 MEL 1720 | 856 150 MEL BNE 1840 2040 | CHK A | BNE | NBP
140 CBR 1555 | 877 125 CBR MEL 1620 1720 | 877 145 MEL LST 1830 1925 | 877 115 LST HBA 1950 2015 | 864 110 HBA MEL 2045 2200 | M/S MEL O.O.H. | MEL | NBQ
150 SYD 1600 | 826 150 SYD ADL 1805 1935 | 845 145 ADL SYD 2010 2230 | SYD | NBR
827 140 ADL SYD 1515 1735 | 889 140 SYD CBR 1915 2005 | 889 80 CBR MEL 2030 2130 | 878 60 MEL CBR 2200 2255 | CBR | NBS
5 125 D CBR 5 1605 | 886 140 CBR SYD 1630 1720 | 887 135 SYD CBR 1805 1855 | 888 100 CBR SYD 1920 2010 | CHK B | SYD | NBT
80 TSV 1545 | 815 150 TSV BNE 1610 1750 | 810 135 BNE TSV 1835 2030 | 810 65 TSV CNS 2055 2140 | CNS | NBU
809 115 MEL HBA 1515 1625 | 816 125 HBA MEL 1650 1805 | 840 150 MEL ADL 1910 1955 | 840 150 ADL PER 2030 2220 | PER | NBV
841 145 DRW ASP 1530 1725 | 841 150 ASP ADL 1755 1945 | 841 125 ADL MEL 2020 2200 | SCHED NO2 E/C | MEL | NBW
. 16 . . 17 . . 18 . . 19 . . 20 . . 21 . . 22 . . 23 . . 24 . . 01 BASE

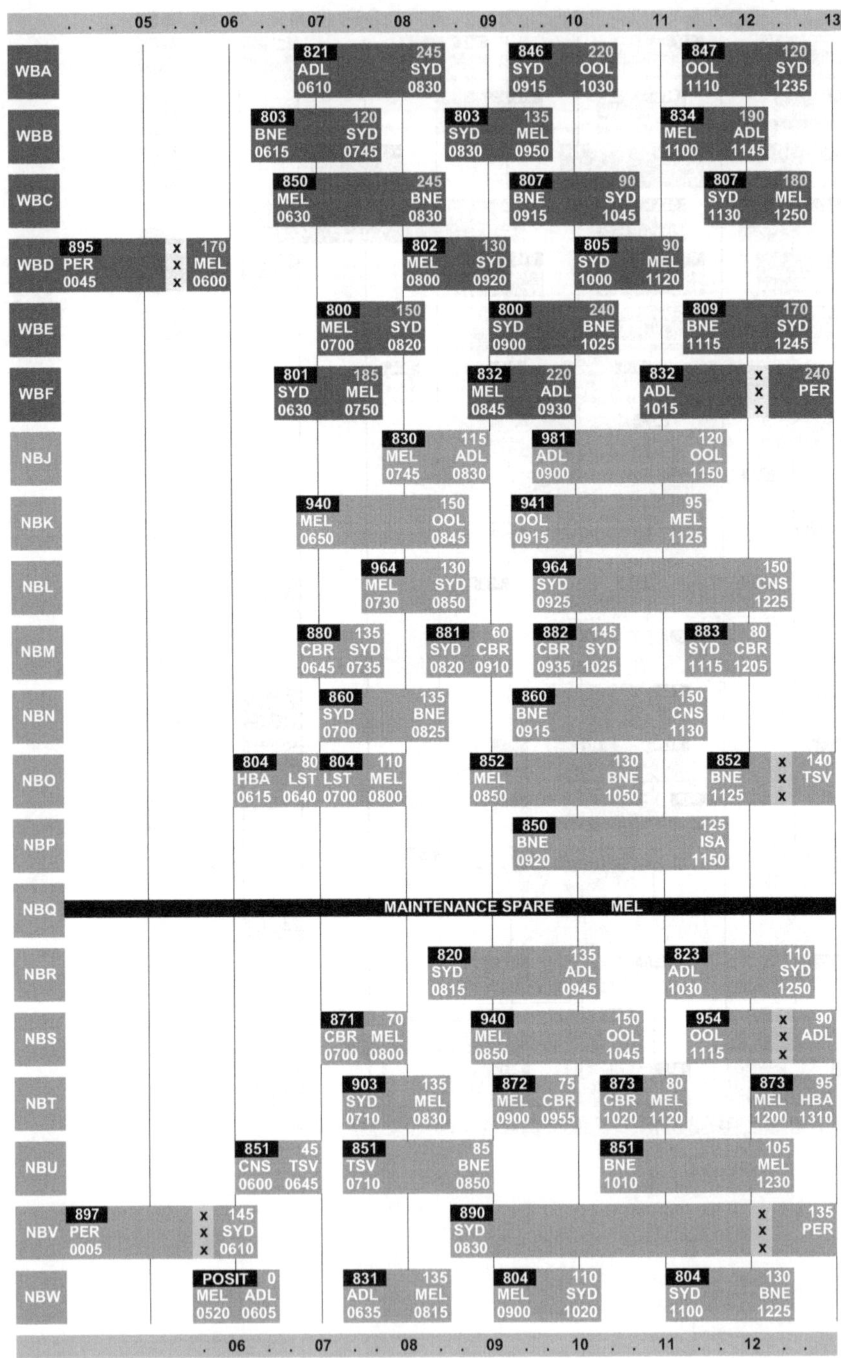

Figure B.2 Domestic utilisation (day 2)

Index